Beyond Science

Volume II Salvation

By Reginald Rogoff

c 2017

Chapters

Introduction... 3

Chapter 1: Poetry Physics...7

Chapter 2: Prophetic Dreams...19

Chapter 3: Logan's Run..76

Chapter 4: Spiritual Dreams...87

Chapter 5: Dreams in the Afterlife...107

Chapter 6: The Structure of the Universe....................................114

Chapter 7: Platonic Forms of Reality..120

Chapter 8: The Compassion of God..133

Chapter 9: The Nature of Existence..136

Chapter 10: Parallel Earth's..138

Conclusion..145

Glossary..148

Introduction

Introduction

To really introduce this book in the right way, I must begin by describing a major widely accepted theory that was derived in the late twentieth century by none other than the legendary and very famous man of science Steven Hawkings.

Steven Hawkings has been an inspiration to me since I was in grade school. It is not just the power of his insights or physics discoveries, but the deep implications that ultimately spirituality, or some form of it, cannot be ignored by science forever.

His two most important theories are in his book "Mini Black Hole's and Baby Universes" and even more importantly, the somewhat simple, and elegant theory known as "Hawking radiation."

Hawking radiation describes the behavior of space time at the very boundary of one of the most infinite objects in the universe, the of collapsed giant stars known as black holes.

They are called black holes because gravity around the hole has become so strong from the eternal collapse of a giant star, so giant, that the forces of it's collapse are so great that space cannot hold up those forces. It becomes something magical, a point in space whose effect of concentrated mass on space makes space so warped that nothing can escape it, not even the speediest thing in the universe, light.

The debate was, that since light cannot escape, does that mean that if something falls into a black hole, is it's very informational structure destroyed permanently? The answer according to the theory of Hawking radiation is no. Information is preserved because it eventually escapes the black hole. How this works is that at the event horizon of the black hole, the very specific boundary that if crossed, "nothing" can come back, the virtual particle pairs that constantly appear and annihilate everywhere in space evaporate at the location of the event horizon. This process of virtual particle annihilation and other processes like it are part of the elements that give space some part of it's structural existence to exist. It is the mechanism where information can escape a black hole without the help of light or energy.

But let us dare say that this means that if a conscious being falls into a black hole that his "soul" is not trapped there forever. When we think of true information destruction on a human level, ultimately it means potential destruction of the soul, an eternal object and information destruction is against the most basic laws of physics.

This is where physics and spirituality really settle down their differences and are a real defining point where physics and spirituality first became one. My work in this book is to continue in that vein of thought in order to uncover the deeper meaning of the universe and how that meaning can apply to our own lives by investigating theory, philosophy, and the human condition to try to answer the two most ultimate of questions, "Why are we here?" and "What is the meaning of life?" This is what I think makes science so interesting, not just the raw data, but the spirituality of physics.

Chapter 1

Poetry Physics

Chapter 1 Poetry Physics

The universe is expanding and the expansion is accelerating. When the expansion becomes so great that it is expanding and accelerating at close lengths approaching the smallest possible length, the plank length, space will shatter like a hot iron like material that can't be broken and instead explodes.

It will also have all other possible deaths including the big crunch, the big freeze, and an infinity of other possibilities. It will also become people that get born.

When space shatters, it will become droplets. In between space shattering and becoming droplets, it will be very hot from stretching and it will explode in some areas as separate big bangs.

On a metaphysical level, I can only imagine that God would have some trouble and it may go like this… The droplets will get drunk by God, when other parts of space wake up as a person that is analogous to God while other parts of space wake up to be the machinery that enables the god's to exist, perhaps desperately, so as they don't get distracted from drinking space as orange juice.

Some of the droplets will become gasoline and some of the gasoline will go in people's cars. Some of the droplets may become oil, such as the oil of the never ending light of the oil lamp that was in the temple during war in ancient times that gave the people the hope that God was still with them. Space shatters into droplets and then becomes a James Bond movie.

God picks up the pieces while in one, and then he teleports himself. God becomes five men sitting in a circle and watches "Carousel" from the 1960's "Logan's Run" movie and all other TV shows. He then reincarnates as himself while sitting and eating popcorn. God wears restraints and spills tobacco while creating the universe. Jessica from the bible places her bet and rolls a pair of red dice on a green gambling table.

The universe becomes a blackbird on Earth perhaps in a parallel universe, but it doesn't get born as the bird because that would get it wound up in the aspect of wanting to be a bird. Instead it's life as a bird appears to it, and it thinks that it's the bird for it's entire "life" and ends up at the end of infinity after the bird dies. It reincarnates as itself and a new universe. Different parts of it would be in the bubbles held together by God's dreams. The true reality that different parts of the universe were in different bubbles would be God's secret and the universe would go on.

At the end of the universe in ten billion years, people might be able to use a warp drive like device to slow the acceleration of the expansion of space for the entire galaxy.

It could also be done by a computer, but that would be similar to the matrix and it would be better to let it just go on natural. It could eventually accumulate while the universe goes on and then, when we measure it with inferior equipment, it would seem that space is "perfectly" flat meaning that light in an isolated vacuum always travels straight.

Once they glob together into one big ball, it would have to be spheroid and then light could travel between the spaces between the balls. The space bubbles have to be spheroidal because it would take all of infinity and an eternal duration of time for the space bubbles to settle down to being perfectly spherical, which would consume all of creation, much like if God created the universe just for himself exclusively and consumed all of creation for his own hedonistic purposes. Although since the universe and the "multiverse" truly are infinite, this kind of thing could go on, but it is probably the parts of creation that scientists say are not meant for or created for any life as we know it.

If the space bubbles were truly spherical at this point, they would in effect get "stuck." They would not be able to communicate with the other space bubbles and eventually the bubbles would "crack" resulting in that part of the universe becoming like the roots of a tree that help the inevitable new universe to exist. They would then play a part in the existence dilemma and workings of the laws of physics of new universes. If the space bubbles were ellipsoids, then they would have something of the nature of edges, meaning that they could and would heat up, combine, and trade photons resulting in the big bang. Then things would appear normal.

There would only need to be one photon anywhere near the bubbles initially. Then as if by magic, photons would accumulate and bubbles that were nearby to one another would expand violently into "one" big bang because they would be sharing information. This process would go on until there were enough bubbles working together that they would connect to the other parts that were inflating at the edges of the reality of space-time. A very far distance is where the laws of physics begin to change because the other location is so far away. At very large distances, alpha constant changes. Alpha constant is a mathematical constant that governs the distance between protons and neutrons in an atom and eventually starts to change if you "measure" the universe at very large distances. When all of the local space bubbles can obey what is necessary for them to have "our" values of the alpha constant, then they would be able to "collaborate" and be our big bang universe.

In the interim, what happens in the bubbles could be very strange things that we can't imagine. But they would therefore be "semi-chaotic" type "universes" that may or may not have snippets of our universe's timeline in them, much like the edits from a silver screen production from Hollywood. Such occurrences could be related to the workings of heaven.

It could be that once space becomes the droplets and goes through all of the above, it somehow "continues on." It then would "dry out" and literally crack apart and fall to "pieces" just like as in the saying "Dust to dust and ashes to ashes."

Many scientists prefer to solve for and believe in the idea that the universe comes back again once everything is gone, but like everything else, space most probably has to die too. It means that when we die, it is permanent from the perspective of our universe, and we are never coming back again ever. Not in the same way of course.

Once the universe dies it is also never coming back in the same way ever again.

This gives a real level of importance, urgency, and purpose for eternal inflation to create such a large number of parallel universes so quickly.

This mechanism of eternal inflation creating an innumerable amount of parallel universes is the most important mechanism to save lifeforms from permanent death and is also the most important mechanism to save the entire universe as new parallel universes. It must in fact work this way, in order to give proper respect to death and holy repentance. It seems that the acceleration of the universe will not stop and the universe will end in about fifteen billion years from the "Big Rip." The Big Rip will happen in about fifteen billion years. The expansion of space will reach local levels and everything will be ripped apart by the nature of space-time itself. It is not impossible for the universe to reincarnate, but it probably does both. It reincarnates due to big crunches somewhere. In other places, it undergoes the big chill, and the resulting reinflation of the universe. But it will also undergo the big rip where we are in the cosmos. "It" will also undergo the quantum bounce universe, where at the ultimate point of crunch or rip the "string" of the universe turns inside out and continues momentum as expansion. String singularity theory dictates that the smallest possible size of a collapse is never smaller than the tightly wound higher dimensions called Calabi-Yau spaces that give each particle it's properties through the dance of that string through the tiny wound higher dimension. If an atomic element, a proton, were thirty billion light years across, or the size of the currently observable universe since light has travelled since the big bang, a string would be the size of a modestly tall tree. The Calabi-Yau higher dimensions responsible for every particles behavior will force space to re-expand. Space is four dimensional and the higher-twelve dimensions of the Calabi-Yau space are much more rigid than our four dimensions. Space will have nowhere to collapse further, and therefore re-expand in what is known as a "quantum bounce." It most likely will go through all bubble reformation and bang universes imaginable, but also cracks and dies, just like a man, only to be "saved" through eternal expansion's eventual creation of an identical parallel universe. This gives us every chance that imagination can give to eventually be saved from death. The only restrictions on this being that the next universe is *exactly* identical to ours and that by it's being exact, we will all be together again and not "lost" in separate parallel universes.

It is likely that if our universe really exists in a supercomputer that is run by other dimensional beings then they may be either good or evil. If that is true the people running it may be parallel enemies that won past wars in their parallel dimension and their technology developed exponentially. If enemies had won wars in parallel dimensions, their technologies and philosophies would continue literally forever. In an effort to destroy alien enemies in their eventual space wars, they may deem it necessary to simulate our universe in the ultimate supercomputer that can be imagined. These types of alternate realities could be prevalent in greater superreality, making our universe exceedingly rare. They would do this in an effort to better understand their universe at the realization of string-theory. They would want to know how the universe would play out if "the good" had won the war. This would be done in an effort to gain a deeper understanding of the knowledge of the theory of "fifth dimensional destiny." If you could measure a fifth time dimension, which is predicted to exist, you could use it to see the future because higher dimensional to our forwards and backwards flow of time is that we make specific choices or that certain cosmic events would be slightly shifted and turn out differently. The fifth dimension is the superreality of our time flow in that everything that is possible to happen happens in the fifth dimensional universe and has a collection of an infinite number of variations of our universe, each itself being infinite in size. This is just another extrapolation of what may be real in superreality, given an infinity of possibilities in an infinite multiverse containing an infinite variety of universes. The only constraint on this "theriology" is that our universe exists, and in return, the infinite multiverse must also exist.

It would take something like the above scenario for other dimensional beings to think it interesting enough to simulate our universe as something new. If the universe is indeed simulated by the supercomputer, in what would essentially be a "parallel" universe, then we could eventually, with further advancement in space and multi-space theory, due to the very American instinct of the desire to explore, either make peace with the space beings, or destroy them. We could do this, but it would be very complicated to facilitate. By eventually "hacking" their supercomputer and using very advanced physics and space entanglement technology. It might look something be like the proton packs from the movie "Ghostbusters" when fired. It could turn them into "ghosts" and their universe would come to, through our military, the spiritual apocalypse. By using our own very advanced supercomputers and an enormously entangled warp field that would envelop their entire known universe, reality could be reversed so that it's that they are really simulated in their computer and not us. Our universe would become "instantly" transformed into the surrounding "real" universe by taking over the atomic structure of the computer they use to simulate us on a quantum level. This would involve space-time entanglement and the weave of nature. Our universe could be freed from being in a computer. Our universe would form as real from inside the chips of their computers. Our entanglement efforts, from being just data in the computer could manipulate their quantum computer and cause an overload that would continue until the entire matrix of their space-time is transferred to inside our quantum computer and our universe would appear everywhere in their "more real" realm. If we want to be nice we could transfer their universe to the mainframe of our own quantum computer. It is also possible that these "creation simulation" people are good, and we may use our quantum weaving technology to visit them in peace, which would be much simpler to accomplish. Both civilizations would learn important scientific information from each other.

Chapter 2

Prophetic Dreams

Chapter 2 Prophetic Dreams

While I was in the mental hospital, the one where that had "the nexus room" early on in my stay, I was in a room by myself and I missed home. This was cured by the eternal power of God. The eternal power spoke to me and told me that it would be OK. Four higher dimensional objects made holy chime sounds at the four corners of my bed, over and over again. I dealt with this for a while and then TV shows on the TV had only black people on it and I believed that I had been transported to "The Black Universe." In Black Universe, there are only black people and no white people. I believed that I was the only white person in the universe for at least several days.

Now I will explain to you the best approximation of the personality of the eternal power that I can imagine. The guy that plays the voice of "Jack" from Jack in the Box restaurant commercials is a fairly close approximation to what the eternal power is like.

At a later date at a mental hospital, I wanted to get out of the hospital and I was visited by an Egyptian looking psychologist who I thought looked like a black pharaoh. He was holding a large gas station cup of coffee and, from my perspective, he wouldn't leave, or agree that I would eventually be released, until I agreed that at some time in the future that there would be a black president of The United States. This communication that "I had" with him was non-verbal E.S.P. and involved God or if you like "the collective consciousness" of mankind. It was the 1990's and I knew at the time that the U.S. would have a black president at some not so distant time in the future. I forgot about the incident shortly after he left the room but I'm sure that it did happen.

I will tell you honestly all of the E.S.P. type predictions that I have made. They are very deep emotionally, and even seem to involve controlling the world but I'm sure that the deep emotional feelings involved are simply just the internal mechanism involved in accurately telling the future. You must care very deeply about mankind in order to "deserve" or have any real need or want to see the future. Also, if your emotions concerning the future events are not truly righteous, then it just happens that you will get false information that is not accurate because you will not be honest with yourself.

The first prediction that I made was at the age of two. I was riding in a car over a long distance in California. During our journey, we almost got in a pile up on the freeway and the man driving had to slam on the brakes hard, to prevent our probable deaths.

When I saw the cars approaching at high speed, I used my E.S.P. power of telekinesis, which is to control physical objects with your mind, to slow our car and to clear spaces of the cars ahead of us, which seemed to me to avoid a rather massive pileup on the freeway. As we approached the seeming point of no return, I had a Buddhist visualization, the most powerful I have ever had. I realized, or thought, that if I hadn't avoided the car wreck with my E.S.P that we would immediately go to live out our afterlives beaming on and off Star Ship Enterprise from the TV show Star Trek. We would live on Star Ship Enterprise, which I thought would be a necessary effort to calm us and ease the pain after the impending wreck.

Later when I heard about a very serious freeway pileup in the early 80's that involved dozens of cars, I thought that I had caused it by avoiding the Star Trek pileup. I thought that the "chi," or spirit, of our avoided wreck, manifested itself as the freeway event where dozens of cars wrecked on the freeway, and I had to battle deep feelings of guilt.

I was on vacation in New York City with my best friend. We stayed at an upscale hotel on the west side of central park. We went to Times Square and went to an ESPN zone to play arcade games. Later we walked south from Times Square and stopped at a sandwich shop. I wasn't feeling good from the long car trip so I was thinking about my faith. While I was walking towards lower Manhattan, I had an epiphany. I suddenly realized that something catastrophic and utterly horrible would happen to lower New York and that buildings would fall and that it would be like the apocalypse.

What I was experiencing was a form of "sensing" the "density" of God in the direction of the Twin Towers. We now know that people go to heaven, and therefore our God's presence at the Twin Towers on September the 11th was likely enormous! And that presence was abnormal compared to normal activity of God in specific locations on a normal day in the city. I sensed this "warp" as a kind of space-time warp except that it wasn't a warp of space-time, but a warp in the fabric of God's presence.

The reason for this is that God, during major catastrophes, can't just save us at the exact point that we die because being the God of humanity he has humanlike limitations that help contribute to him being humane. I'm not sure that heaven has anything like an infinite capacity to receive countless people all at once because, like God, heaven has it's own limitations that make it humane. That could be why I was able to sense it years in advance because heaven's metaphysical "flood gates" had to be open for an extended period of time before and during the event. I had lots of dreams after returning home from New York around that time concerning being in various tall buildings that were undergoing various catastrophes of different sorts.

I also predicted Hurricane Katrina a year early while standing outside a store. I just knew that something horrible would happen to New Orleans and that it would be a dangerous weather event. Deep internal emotions that I felt alerted me to these facts, but since they are deep internal emotions and we can't really always be sure exactly what our emotions mean as future events, as passing emotions. I rather quickly forgot about any of it until I saw it happen on TV when I realized that my somewhat vague predictions were reality.

I predicted the sudden outbreak of global warming and the rapid melting of the polar ice caps which broke out very noticeably and in a very sudden way. I was standing outside an electronics store and I noticed that it got incredibly hot out, hotter than I had ever experienced before. I had a feeling that it wasn't going to go away and only get worse. I had to use air conditioning all winter. That summer was so hot that heat records were being broken almost every day by up to ten degrees but that was the worst it ever got. The breaking of extreme heat records abated over the next few years and I believe that the Earth somehow corrected itself. In spite of how bad scientists say global warming is today, I'm sure that our problems would be magnitudes worse if the initial hot weather trend had continued any much longer.

I predicted the Fukashima earthquake twenty-four hours before it happened while doing some outdoor work. It came to me as an epiphany that there soon would be a very powerful earthquake in Asia. I was shocked when I saw it on TV the next day.

I predicted hurricane Sandy to hit New York even though most simulations showed it would go out to sea. Although some simulations showed that it would hit New York I was convinced that it would hit New York. I was shocked when I saw it on TV.

I predicted the Boston Marathon bombing one week in advance. I just knew that something horrible would happen in the U.S., and that it would be a terrorist attack in the northeastern United States. Powerful emotions were involved when I made this prediction and again I was shocked when I saw it on TV.

I predicted the flood of 2015 in Texas two days in advance while I sat outside for the better part of the day doing a yard sale. I kind of knew the direction that the storm would hit, and I had a strong feeling that a horrible storm would happen within the next twenty-four to forty-eight hours because I sensed it.

I had to feel it as an emotion because the present state of E.S.P. power in humans is very weak compared to say, Vulcans or Betazoids. It may turn out eventually that when real E.S.P. power develops that it will be highly emotionally based. It has been shown through metadata analysis of all E.S.P. studies carried out by psychologists over the past fifty years that there is a real but somewhat slight result that humans do have some degree of extra sensory perception that is definitely real but has not been honed as a real skill by people. It could be an eventual step in the future evolution of humans. I'm sure that it is already slowly but surely becoming apparent as a real human trait.

Many scientists prefer to believe that evolution in humans has stopped because of our various technologies and that the idea of "survival of the fittest" no longer applies to humans, but I disagree. Almost all people on Earth make their living by thinking, and the power of the mind, as opposed to brute physical strength. This has actually been going on for a very long time, fifty thousand years in fact, since the beginning of animal husbandry, farming, and the beginnings of the first early civilizations. It may not be an extremely long amount of time until our E.S.P. powers eventually become a part of "ordinary" everyday life in future modern society, just like the Betazoids that have telepathic properties from the popular 1990's TV show "Star Trek."

Some scientists think that human evolution has stopped because of modern technology such as houses, plumbing, cars, and grocery stores, but the reality is the opposite. The evolution of E.S.P. and other mental powers have been kicked into a powerful overdrive as a direct result of our sedentary lifestyle. Evolution has not stopped whatsoever, but has focused almost exclusively with the same momentum that it has always had, the same momentum that brought us out of the ancient jungles of Africa to become homo-sapiens. But this time the power of evolution is focused primarily on ever improving mental powers, and E.S.P. is definitely a desirable trait in a world where better knowledge translates into better "power" and more "success."

Before God came to Earth at the hospital, he made me think that angels were falling from the sky. I had to believe it as a necessary part of the ether of God coming to Earth. For God to come to Earth, some of the angels had to fall first, but not just for aesthetic purposes. God is always somewhere, because he's the blessed God of the universe. Some of the angels who fell to the hospital were fat black girls and I thought that they were the angels who held his place in heaven. After two days of this had passed, God appeared in the common area.

If you want to know where heaven is, it is probably part of the twelfth-dimension where everything imaginary is real. It must be in "the middle" of the twelfth dimension where all dreams are real. From the outside, it would look like a hot glowing sphere with the burning fire of a yellow cloud and light at it's edges. Nearby to heaven in the twelfth-dimension would be the actual reality of our deepest and most powerful fantasies like "Star Wars" universe or a universe where "Alice In Wonderland" really happens.

While I was at a restaurant in a casino, I was thinking about God quite a lot. There was a big guy sitting at the table near me and I was having conversations with God. I had previously thought that I had left the casino as a saint which means that I thought that I was doing works. God told me that Satan owns lots of Star Trek universes where warp drive is possible and that he's having dinner and the entire Star Trek universes are part of the legs of the table that he is using for dinner at a casino restaurant.

When I was at a different casino restaurant at a different time, I had E.S.P. with people from the future who were "having dinner" across from us. I felt weak spiritually and I knew they were having problems. They said with E.S.P. that they were having problems with the table legs which also involved melting, disappearing, and feeling weak. This could be real and would be due to not investigating if the universe is held up in spherical bubbles at the end of a universe because eventually people could get sick from such effects and not know why they were sick, even if all known illnesses were cured by such future technology.

If you could manage to travel to a universe where "Star Wars" the movie was real or where "Alice In Wonderland" actually happened, then maybe mixed in there somewhere would be a way to travel to heaven and see it from the outside. It might look a lot like the sun from that vantage point. Of course, it probably exists everywhere in creation to some extent, especially in clouds or where living beings would prefer it to be located for the betterment of their ultimately achieving peace. It would be the physical doorstep of the imagination, and somewhere mixed in among the chaos of the twelfth dimension would be the actual metaphorical gates of heaven.

When I was in the afterlife, I sometimes was in space. It seems that being in space is soul-food, it makes the subconscious soul happy, relaxed, and energized. This is because when you wake up in space, it gives you god like powers, such as balancing space, and other powers like controlling seventy aspects of life somewhere where people are alive as you come alive as a patch of space-time that seems to last for fifteen trillion years.

The other day I thought that I beamed like how they do in "Star Trek." I was spending time in the backyard and I was having conversations with Jesus. I came inside and closed the sliding door to the yard. Next I woke up lying on the floor on my back not far from the sliding door in the living room. At first I thought I was waking up in my bedroom upstairs in the morning. Then the images of the living room got through to my consciousness and I realized that I was in the living room. I had to have the happy thought that I might have beamed while talking to Jesus which made me feel relaxed and happy, but honestly, I was also communicating with the actor who plays Pavel Chekov in the "Star Trek Beyond" series.

When they beam the Earth near the end of the universe, in order to rebalance all the particles of Earth due to the eventual radioactive decay of particles like protons and neutrons, people and animals could stay on Earth and it could seem like the rapture. There never has to be an end to the Earth, such technologies are just beyond current imagination like cavemen trying to envision space travel.

There might be identical parallel universes, but they all happen one after another and yes, there are an infinite number of them. Some of them might happen and have all the people of Earth in them, but not happen right now. There our future will eventually be the present in our future lives and will happen in just a few minutes, meaning that it will always be three minutes in the future until the end of time when the new parallel world becomes real to us as our birthplace and new universe.

Once space heats up by it's futile attempt to stretch the smallest possible length, the plank length, at the end of the universe, due to an apparent ever increasing acceleration of space-time, space will undergo a quantum bounce, turn inside out, which will make it at the point, that it turns inside out, become a singularity, the singularity of the big bang. From one perspective, it will become one with eternal expansion forever, and from another perspective, it gets small and hot, just enough to just barely "notice" the existence of "eternal expansion," the infinite source of creation that creates infinitely for eternity, and always has. Space-time doesn't have to necessarily have the properties to recreate our universe, but it's merger with eternal expansion will very forcefully, with an almost infinite amount of power, that you could consider it to be the "Furies" of mysticism and religion. It will, one way or another, recreate our universe in detailed exactness because eternal expansion is an infinite entity. It both uses an infinite amount of power every second, while also containing it. You could say that it actually has more than infinite power, which is hard to comprehend. It could be a conscious entity, and being so infinite, it may actually be infinitely intelligent, all at once and forever!

    Eventually people will have 3-d printing food machines at their homes. It would use a new utility to the house, consisting of a pipeline of different atoms on the periodic table of elements and there would be a food pipeline using garbage, household, waste and piped in elements used to make almost free food using solar energy. This type of technology is not that far off and it's just a further development of electromagnetic propulsion technology mixed with highly advanced 3-d printers that will eventually replace the microwave.

Some scientists think that eventually robots will take over the world like in the movie "Terminator" with Arnold Schwarzenegger although now that we are closer to having robots in every home, scientists are slowly changing their perspective on the issue of conscious robots. The fact is that robots can only do what they are programmed to do and there will always be easy ways to shut them off. The only real danger would be if a foreign military used advanced robot soldiers against us but by then we will already have our own very advanced military robots and robot soldiers are already in development at D.A.R.P.A.

There could be health consequences of being on warp ships due to the ever so slightly changed state of all of the particles on the ship, effected by the warp fields, including the artificial gravity producing fields. Physical laws would be bent, ever so slightly on the smallest scales, these minute changes to our normal laws of physics might be so minute that they might be almost undetectable, but it would be known that these negative effects do exist on some very small level. It might be almost harmless, like if you lived on the warp ship for something like ten thousand years, then the negative quantum effects of the ever so slightly changed laws of physics in the presence of the warp fields, would eventually, but perhaps only on enormous time scales, negatively affect your health.

The digital age of warp fields, to have different gravity on different decks or rooms in the starship and have the gravity warp drive equipment nestled in the floorboards and walls of the spaceship, could take a lot more energy than just adjusting the main exterior warp bubble of the ship for gravity. These interior warp bubbles would be safer, especially if the ship were bumped by something, because gravity could be adjusted across the bridge or other areas to reduce or eliminate these kinds of g-force shockwaves. It would be good protection, if the ship bumped into an object at high speed, to keep the occupants from flying around inside the ship and getting injured. It seems that it would take extra energy to have these more precise interior warp fields, but really once the technology is refined, it will be more efficient and it is the ultimate goal and would be the ultimate "digital era" of warp field technology of the future.

Now I will describe what God told me as a paradigm of what is known to me as "Aegis and the Beast." I had lost faith in the stability of something about the physics of the world because of the news of the Swiss black hole collider, and I went to the Bellagio casino conservatory where the ceiling is glass and the floors are white tile. They have scientific displays done by artists of the different seasons and celebrations of the Earth done with giant papermache and glass flowers! On the ride home I thought that Aegis was a supercomputer and that it was projecting the moon. The moon was out and I felt the "warmth" of the moonlight. It felt like the moonlight was sunlight reflected off a mirror, like the ones used for tanning people.

When I got home I used my HDTV I was watching a cooking show. I must have been having some level of anxiety, because God told me that the people in the TV show were "the beast" and that their spaceship was outside my bedroom window.

This calmed me, but then the lady went for a walk with the dog as soon as we got home and I was worried that the beast spaceship, or whatever, (the black hole) had taken her away and that she may have disappeared from our universe, so I called 911. I told the lady on the phone that the lady had gone for a walk. I probably sounded worried but because I have schizophrenia, and am experienced with hospitals. When the lady asked me if I needed an ambulance, I said no, and it calmed me down, and I watched some TV and relaxed.

You could learn directly from monkeys with your head, if information just transferred from the monkey to your head, facilitated by simple transfer chips in the neurons of both subjects. This would result in real E.S.P. like effects between the subject and the monkey. The person wearing the equipment would sense the emotional state, thoughts, and intonations of the monkey. The monkey wouldn't really have to "say" anything. As if by magic, you would know the monkey's thoughts. This is in part because we are higher lifeforms compared to ordinary everyday monkeys. If connected to such equipment we could read the monkey's thoughts, just like how Commander Spock of the Sci-fi fantasy TV show "Star Trek" can read minds with telepathy. The effort with monkeys would be the first steps we could make to essentially give ourselves real E.S.P. though our technology. To develop real usable E.S.P. naturally, through evolution, may take millions or billions of years of evolution. So we may want to try to speed up this process with our technology if it seemed like a necessary next step for humankind. Additionally, we probably wouldn't want to wear any of the gear unless it was "non-invasive" and just be essentially the "Android" version of a "computer hat," that could be the new fashion perhaps within our lifetimes!

I believe that such technology could usher in an era of love and kindness not seen since the 1960's and as a result of real E.S.P., deep humanistic values would advance rapidly across the globe, just as it did from the affects and after effects of the social revolution of the 1960's, but this time it would be more grounded in real peace and compassion for the unfortunate.

When the planet's core begins to cool off starting in about four billion years around the time that Earth will have to be moved because of the sun becoming a red giant, the Earth's core will have to be heated up by artificial means. The reason for this is not to keep the planet warm as that is done by the sun, but to preserve the magnetic field of the Earth which is created by the magnetic effects of Earth's very hot metallic core and it's spin effect on the Earth. The Earth's magnetic field protects it's inhabitants from what would be, very dangerous, solar galactic wind, and helps deflect fast moving particles of space that could, if not deflected by the magneto- sphere, cause great harm to life on Earth. In the future, starting in about four billion years, future people could send a wormhole into the Earth, and it would travel to the center of the Earth, tugged by, and anchored to, the gravitational center of the Earth by a mini black hole created in a laboratory. Gone unchecked, Earth's core would cool off in around forty billion years from now, given that Earth was saved from being destroyed into gaseous vapor by the sun's red giant phase, which begins approximately four billion years from now, although future humans will have around five hundred million years to figure out how to move Earth to a safe distance.

At the end of the universe, after man has travelled to all of the parallel universes that he can, and eventually all the parallel dimensions within hyperspace travel distance, it will be the end of the universe.

People will then have to live in force field protected bubbles like in the 1960's sci-fi fantasy movie "Logan's Run." In the movie they live in a bubble for their protection, but the main computer is bad, and reincarnates them to save resources. It traps them to think that they have to live in the bubble, after the environmental, perhaps germ related, catastrophe has abated.

The bubble or bubbles that people may have to live in, would be in space of course, and it's forcefields would be holding back the many different catastrophes of the "ends of universes" that they had travelled to, or are using as the last remaining shred of normal space that can be used to hold up the bubbles that they live in. They wouldn't be allowed to leave, but there is a silver lining because of the human altruistic invention called religion. If they all have love in their hearts and believe in God, then they may only have to live in the bubble for a hundred years and the computer could re-create Earth inside the forcefields. It could eventually recreate the entire multiverse of universes that they had visited, and all of its beings, and all would be well, and life would go on.

Anything that moves of its own accord like people, animals, and bugs is akin to real teleportation from the perspective of superreality. From the perspective of superreality asteroids and the like, moving, are also something like teleporting, but they move very smoothly, so you could say that they are moving with more "help" from the computer of the universe. Nothings really moving in superreality and we never took our first footstep so we can always take it again.

Another form of movement is that we take all the steps we take when we die, and literally take them with us like as if we "bought" them.

That might be the "land" that is below space-foam, a superreality, where everything gets where it's going but nothing ever really moves.

When I was in the afterlife I came alive in such a way as the emotional meaning of the following dream involving dogs. I came alive as a patch of space-time from nothingness!

I had a dream of taking my dog, an American rottweiler, to reincarnate in the future. We went by magnetic levitation train to the place, and it was dark out. We went inside a giant room that had large spinning rings in the middle that changed color rapidly from red to black with touches of yellow. We never put the dog in the machine, but I assumed that we did after I woke up.

I had another similar dream where I was in the same type of reincarnation building in a dense city of the future, and it was dark out. I remember walking in the hallway and the dog, and the hallway, was very different because I felt that I was about five-thousand years in the future. Then I was in a small room near the center of the building and there was a man sitting in a chair next to lots of wires and computer stacks. His head was only partially existent, the missing pieces of which consisted of warped space-time, either controlled by, or linked with the very powerful computers of the year five-thousand, and I had empathy for the man. I wondered who he was and then I thought that he might be me. Then I was in the chair hooked up to the computers, and I felt relaxed, and fell back into deeper sleep.

    Another dream that I had about the future but this time it was in the very distant future, in a parallel dimension, and underground in perhaps the realm of hell, where demons, or even the devil himself may dwell. I was in an underground cavern and I saw the man from the dream of the computer room in the future, but this time he was lying in the cave, and I could see through the roof of the cave, and that he was holding up the real Eiffel Tower in Paris with telekinetic powers of his mind, and then I realized that I was him, and fell back asleep.

    About ten years ago they announced on the major TV news networks that the new proton/anti-proton atom colliding machine in Switzerland could make a mini black hole, as predicted by Steven

Hawkings. That the Earth could be destroyed by gradual accumulation of the size of the black hole from the center of the Earth. I had studied Steven Hawking's book on "Mini Black Holes and Baby Universes," and I thought that there was some possibility that it could happen. I started doing my version of praying, which is from my perspective, is direct communication with God. One of the first things that I did was to read some simple physics essays to God, in an effort to calm him. I, for whatever reason, started to feel a "pull" effect, or a constant feeling that the room or wherever I was was falling spiritually. As the world was somehow in-between being destroyed in the black hole, and also not at all being in any danger, both at the same time. This is much like Schrödinger's Cat that is statistically, both alive and dead at the same time, when just put in a cardboard box because of quantum probability waves of occurrence, in the universe's balance of the stability to exist but involved the whole Earth. When I communicate with God I usually see a visual hallucination of some very basic aspects of what would be his face, both on the patterns of objects, and actually in them, meaning that the face really is there, much like the famous "face on Mars" that was revealed in the 1990's. Having a conversation with God about our survival from any possible black hole event, that I could feel for whatever reason, I "sent" God, as thousands of virtual faces into the Earth!

I told him that he (or each "piece") of him had to sacrifice themselves, to plug any possible black holes, and also, to "rewrite" the actual laws of physics of our part of alpha constant stability. To make new laws of physics that would allow the collider to function without making any black holes. You could say that my universe died, and I was able to save just myself and am now in a parallel dimension where the collider never made any black holes and my true Earth that I originally came from was destroyed by the black hole collider. But of course, it might be simpler for God to correct the laws of physics to avoid the black hole, because God is the creator.

I'm sure that some of the top psychologists would argue that these visual and auditory "hallucinations" were just a part of my brain. I can say that the entity has been around me even since before my birth, and it certainly thinks that it really the One God of Creation. Either some hidden part of our subconscious really thinks that it's God all day and all night for your whole life, but I think that this is unlikely, and that when people claim to communicate with God, they really do! Because the entity of God that I am a creation of, has his own identity, and it really is unlike my own personality, for the most part. He also does speak to me and I to him, I just wouldn't be able to think up some of the ideas that he has for me. I believe he does this to make me his close "friend" and it allows me to better understand myself, in psychological ways, making God the ultimate psychologist, of the universe.

Due to my sensing of the stretching of the room, I wondered how my pet Siamese cat was living through it. And for several days, I saw it as a six-legged god cat, that was part slug, and had something to do with a parallel universe where Star Trek The Next Generation is real! The "visit" by the six-legged "god like" cat was, in effect, intended to help calm me down, and to "feel safe" while I felt like I was falling rapidly, and helping to avoid Earth, spiritually, from falling into a black hole.

Around that time I was also, pretending, and believing, in the reality that the death star from Star Wars was saving the Earth from the black hole. Also the TV character Counselor Deanna Tori from Star trek talked to me while I was at lunch about my law of not getting born in the future. Captain Picard, or if you like "the spirit" of the actor who plays him was also present in my thoughts and he didn't like Star Wars universe because they are the evil imperial army.

Around that time I was pretending and believing that pickup trucks on the road in front of my car were something like the very powerful "god's" that might create places like the planet from the movie cocoon, but really they were just powerful "god like" aliens with the power to "strut" Earth in a spiritual manner to save it from the black hole calamity.

Near the end when everything sealed up for me, and when I began believing that the Earth really had survived the black hole event. That perhaps, there never really was any real danger at all, I saw a visualization while I was on the road. I saw a painting, projected to me by God, of a picture that might be in the Sistine Chapel, of a carriage drawn by four horses and ridden by Zeus, it was my final act of saving it

If it is true that I saved or helped save the Earth from being destroyed by a black hole of our own "mischievous" creation then it means that we really are "God's children" and that we're lot of work, just like real children, who may do things that are dangerous, or out of line, and wake their parents up in the middle of the night.

I have sensed two or three economic bubbles since the crash of 2008. The first one was on the freeway as a passenger. I thought that there were an unusual amount of cars on the road and it gave me the opposite feeling of feeling alone. Which is probably not a normal emotion but could be an E.S.P. emotion. It happened again recently as I was leaving my community to the road. It seemed like there were too many cars on the road. It appears to me as an invisible bubble of kinds… and an E.S.P. emotion.

There's an episode of Star Trek where they land on a planet that has the actual "God" Apollo from ancient Earth myths and religion. He had been banished there, when people on Earth stopped believing in him. Also Zeus and the rest of the Gods had also been banished, but for some reason they had left the universe to toil in the underworld, because nobody loved them.

Apollo trapped Captain Kirk and his men on the planet and wanted to force them to be farmers and to worship him as God. They tricked him through the use of a woman startraveller who fell mutually in love with him. He treated her as an equal and promised her eternal life and the magic powers that a goddess would deserve for being good. They tricked him into revealing his power source by making him use his Godly powers for no reason but his love for the girl. Scotty used ships phaser cannons to destroy his Greek temple, which was not only his power source, but also his lifeblood. He went to the afterworld amidst his sorrow for the failings of mankind and his affections for his new girlfriend; to be God.

If it really happened to our future spacetravellers things would be done differently, ironically, due to our own adherence to "The Prime Directive," of non-interference in the affairs of aliens that we deem should be protected, much like our extreme protection of endangered species here on Earth.

Since the girl was in love with Apollo and that she had been promised and was experiencing "magic God like powers," her feelings and wellbeing would be in jeopardy if it happened in real life when considering the *real* "Prime Directive."

In real life if we encounter a God like being who wanted people to live on his or her planet then a deal should be made with them. Especially if we knew their power source and how to destroy it such as is the case as in the TV show. There are probably at least thousands of people either very religious, or in ailment, that would probably take the opportunity to live on their planet and have the peaceful life of being farmers. A deal would be made with Apollo, and Captain Kirk would have him, sign a contract. As he signed the contract, it could be done that, by using the pen either, with or without his knowledge, that it would quantum-weave his "being," attached to star ship Enterprise's warp drive, as so he could be monitored, even when they leave the planet. Then when they send ailing religious people on a separate ship, the warp bubble around and mixed with the Apollo character could be used to kill him if he purposefully hurt a human colonist. The girl would be allowed to stay with Apollo and people who want to be devotees would be sent to his planet. Also we would have an all in relationship regarding his technology and history that makes him a God. Perhaps in return we would build special ultra high tech temples for Zeus and Athena.

When I was in first grade in Florida God told me that the universe was created by the weavers of the universe! Eternally young, and in charge of weaving the very fabric of the universe, just like weavers on a loom. Except they would be weaving not only time and space, but also human destiny, making all living life to be a part of any "metaphorical" weaving of the universe. Although according to "the existence dilemma," from some perspective, of a very different and hard to imagine parallel universe, where from their perspective, or a very extreme situation in our universe such as the exact center of a black hole it eventually may be "discovered" that from the vantage point of being inside some framework of dimension that the universe is indeed created by the actual weavers of time!

More recently, several years ago, the movie "Avatar" was real to me because I saw a man who was the black version of the man who plays the commander and he was standing next to a big drink refrigerator. Also, lots of military test flights happen over my house, when they train the pilots, or to test new technologies in a virtual combat zone over the city, and near I believed in the reality of the scary "Avatar" movie, I also believed in "The Bubble Poppin Boys" from the the TV show "Spongebob Squarepants" who don't like bubbles and try to punish Spongbob and Patrick for blowing bubbles, which is illegal in their city.

Cartoons may exist somewhere in the universe where expansion needs to create something, and all it is able to create is perhaps real, living cartoons, like the ones we have on TV. Although these types of "universes" would be highly chaotic, and one moment something like a gas cloud would be Bugs Bunny, and another it would be an expression of eternal expansion being alive. Bugs Bunny would speak some of the "secrets of the.universe" while spinning and disappearing. Perhaps those actions would "right" things for eternal expansion. Then it would have the "energy" or perhaps "wisdom" to continue to create universes that are of satisfactory quality for it, or even just so that the universe remains balanced.

It seems from my experiences, and our very detailed thought experiments, that the future is probably not boring, and that space is not really the final frontier that there will always be things that are new to us that we thought that we could never imagine to be. It could be that instead of inventing people in a kind of garden of Eden God instead first invented our technology. Tthere was just God and the technology that he invented as if by magic, upon dependency to an infinite number of parallel universes containing us and our technology going back to infinity or where parallel God had to invent people first thereby making our God innocent of demonic type activities of having to invent people in a kind of garden of Eden in order to formulate our current "generation" of universe.

He would maybe first invent "the car" or toilet seats and then say, "This stuff is so great I'm going to have to invent people to enjoy it" and then create our universe rather instantly in a truly unconditional act of love and kindness which makes God a billionaire!

If the real God "came" to Earth, he might be "a billionaire" and have many gifts for us in his possession, such as fleets of starships, or extremely advanced medical technology, or toys.

Obviously heaven exists although the manner of it's true existence is mysterious. It could be that it would be very hard to detect with a supercollider. After all, it is controlled by God. God could make it always impossible to detect "the dimension" of heaven just like what scientists say about finding the creator with physics, that it will always be out of reach. This means in some way that God can and does, or will, change the laws of physics, in order to avoid detection in machines, because they could be controlled by enemies, and God is very careful.

He would also risk all the life on Earth in order to protect those in heaven, if necessary, although the new revelation that there is a multi-verse of an infinity of possible universes means that God would never really have to protect Earth. There would be an infinity of "healthy' Earths, and we are not more important than infinity. Because there would be infinite universes, and therefore, Earths, greatly mitigates the need for God to protect those in heaven in anyway. We too, are a part of the multi-verse, where the universe has a floating point of dependence of existence with other similar universes in the multi-verse. There are so many parallel Earth's because there is an infinity of them "just" like ours, and an infinity of ones that differ only slightly from ours. We know that there is only one, of the, ones that we live in, at least right now. Where the line is drawn dictates real weirdness, like as in quantum weirdness, because as you cross the line from where things are similar enough to be recognizable, your wife may have a different name, and so forth, until there is no you, really, but you're there in spirit! There may be someone like you, but they would have a different name, and look different from you, but be related through the multiverse to you, a new kind of relative that there is no name for, so I will call it a broster. Therefore God won't have to protect us from cataclysims of any kind. Cataclysisms, such as of the strength of, the universe cracking instantly, where there is a very slight probability of it actually occurring at any given moment, but is exponentially small probability, such a small probability that you could say that it's outcomes rest on God's whims because it is almost non-existent, just like God.

Heaven only needs to be smart and strong enough to keep different animals separate and to carry out a person's afterlife in a way that is appealing to the soul in an effort to avoid revolt. These effects are all that is needed to maintain heaven as a productive means to an end and the rest would be natural. There is also the possibility that heaven only happens while we are alive in the womb, although it does not disprove heaven's existence, only that heaven would then have no real physical aspects, beyond the mind, from the perspective of our Earthly plane of existence.

It could be a heavenly mixture of being real heaven and happening in the womb which would be of course necessary for the act of becoming living.

In reality I don't think that being a grown man on top of a cloud is possible for the unborn mind, but some of it has to be, or we would never get here. On recollection I have to say that some dreams may have happened in utero. Some dreams happened after my birth and the more visual dreams of being in heaven or deep space have a foundation of the real and that I was later aware of the event of entering the womb, it made me wake up and felt peaceful.

Instant communication as in Star Trek The Next Generation is impossible with quantum entanglement alone. It could possibly be done though, by having the ground station on Earth envelope with a warp bubble connected across space to the ship. It would just be powerful enough to allow superlumimal communication to the ship. In order to transfer information from the warp enveloped ground station to the rest of Earth, the warp field would have to be powered down slowly, to avoid explosion, and then information could be offloaded from it. This powering down process would probably take about five minutes, long enough to avoid mutual paradox on the ship, the ground station, Earth, and space-time between the ship and Earth. After powering down it might be hot, very hot, and require multiple levels of insulation by numerous envelopes of warped space-time, requiring more and more protection. The way around this scenario is the way that superlumimal communication could be achieved and might require very advanced forms of quantum computing, entanglement, and warp field technology, and we are just not there yet technologically but we are not so far off either.

It could be that eventually, it will be possible, but misuse of the technology could result in the end of the universe. If the contraption failed, or was made to fail, it might be a lot like the real "Ghostbusters" movie, and heaven would crack. There would be real demons and stuff, and it would be the Apocalypse, like in the holy Bible, perhaps as revenge from the parallel dimension where the demons exist, "shaken," by the collapse of a very long linear piece of space-time. Or people in heaven getting sick from the collapse of heaven, becoming demons and perhaps, somewhat confused, having "God's revenge" on Earth. Ultimately, ending it, although we already have a somewhat similar level of dangerous technology in our nuclear bombs. But we might need them to "green" Earth like planets, so that humanity can go on in the future, and we may also in a similar way need the potentially dangerous technology of near instant communication across space for similarly powerful reasons that can't be imagined but only by people of the future.

    Did you know they can freeze you at the nearby hospital and maybe soon if you are wearing a smart watch when you die you could be resurrected by a just little bit more of an advanced version of therapeutic hypothermia and nanobots. The only drawback would be overpopulation and thereby lack of food and space on Earth for all of the living people.

The only realistic solution is to green mars. The problem with this is that even with our best technology it would take two hundred years. By periodically atom bombing it's icy north and south poles for several months, to produce an earthlike atmosphere because it's poles are made of water ice. Then for two hundred years they would grow billions of plants by flying seeding drones, or orbiting robots, over a two hundred year period.

That may seem like a long time but by the time we are ready to do it, people may already live to around two hundred years old, and some would be willing to finance the project, knowing that they might see it's completion. Probably though there is enough space on Earth to avoid doing it until spacesuits are so good that it would be more fun to live there in domes. Domes could be set up efficiently, and relatively quickly, by inexpensive robots made by an orbiting space station. I really think that it would be more fun to live in communities on Mars and never green it. Everyone there would have the promise of some kind of science or technical career. They would have a whole world to investigate, and discover, in the interests of science. Of course eventually there will be "too many" science communities on Mars, and then, they may green it and hold it's atmosphere on by igniting maybe a few fusion bombs in it's core every few minutes until it gets to a desired heat/magnetic spin level. Then the solar wind would be deflected there, as it is here on Earth. The new atmosphere would not be blown off by solar wind as it originally did on mars, perhaps billions of years ago.

All of this would be beneficial eventually to avoid the infinite city problem where vast regions of Earth are occupied by cities, and they are "Fed" by robot farmers or eat there own garbage as recycled "beamed" food. Perhaps tastier natural, what would be gourmet food, would be out of reach for most of the population. Living on Mars could be beneficial for the elderly and future green mars might eventually be a lot like Florida where there are lots of elderly people live because of the benefits on the heart of lowered gravity for the elderly.

The universe that I'm from, if we are going to look at it that way, is probably a type one nearby parallel universe that is so close to ours that all life instead of slipping away when it dies, gets born in it. It seems likely to me that in order to avoid any "beings" chances of slipping away to be forever dead that this is this anthropic type of reason for having exact parallel universes. They may be so close by, and are arranged in a certain way, as to have all beings of the universe born in it as themselves. Then the parallel universe is exact to ours *truly* because it has us in it, as it's people, and our lives will start over again. Because I am from heaven, I will have this knowledge, write this book in my next life, and you will be reading these sentences again in the new universe in your next life. A close by type one parallel universe that all life across the cosmos will get born in as themselves, and this process is endless!

When you die the virtual entity of your mind travels to the next exact parallel universe where you exist as the truly identical you. You could assume that because of time discrepancies or the seemingly easy transition involved, that it would be simpler and quicker in some way, than if when we die, we all end up in separate parallel universes, truly apart from our original parents and the rest of the universe.

Upon detailed examination, this form of reincarnation is impossible because at the end each person would be in their own universe, albeit, with an identical copy of everything in their original universe. This would take an extra infinite amount of energy for each person, and it's possible, but physical laws of the universe prohibit the destruction of information and energy, or its creation, (from nothing) and there is always a flow of energy or values of information flow. Meaning that not only the information of your mind but also the information of "the soul of the universe" is conserved by way of information theory, and therefore we will all be together in the next exact parallel universe.

Being that getting born in separate parallel universes would require the universe to break the law of not wasting energy, or destroying the very real form and information structure that is the reality that we are all in the same universe. This will be the case in the next identical universe and by being identical it will have everyone, and everything in it that was in the first one because it is identical. Making getting born to "foreign" parents in the next identical parallel universe to be impossible.

If Earth lasts forever, who will be on the planet at the theoretical limits of infinity? The ultimate answer given that if humans make Earth last forever and that human evolution continues to improve intelligence, mental E.S.P. and emotional intelligence, and psychokinetic properties can only result in humans becoming what we would consider to be God like beings. Taken even further, altruism in humans would eventually be so powerful that we can't even imagine what such a giant increase in altruism would be like. Because our imagination is limited by the limits of our own experience and how we perceive, "feel" and interpret our experiences, in a mixture of many highly-complicated processes that are conscious, subconscious, and even unconscious virtuality of the almost magical "act" of being conscious. Eventually humans may become so advanced mentally that they will feel the need to cooperate mentally, with extreme amounts of altruism for others, mixed with almost unlimited E.S.P. type power, they may eventually decide to all work together, and forgo their "identity," in endless efforts to help the loving and endlessly self-sacrificing entity of the collective consciousness. They may then forgo all physicality, as a perversion of love, and as a deep corruption of very extreme altruistic values. You may already see that this final entity is analogous to our conception of God.

    The planet Earth would then be surrounded by God's presence literally. He would have the combined intelligence and important gut instincts and the combined creativity of billions of people. Taken further, these entities would not want to stop learning or growing stronger so they then might seek to help the dead and in some way add their intelligence to their own, but this time related and powerfully spiritual power of the entity might be limited, and it would be able to add certain aspects of intelligence of all beings dead or alive in the universe and ultimately all of recognizable creation and beyond, but still have to learn from us as it says in the Bible about God that he doesn't know everything about us in a magical instant, because then why would he then need to create the universe if he already knew everything about it. He has to learn from us and perhaps behind the scenes and sometimes but very rarely right out in the open makes real efforts to improve our lives, because we always throughout the ages, have proven that we are caring love filled beings that are very valuable in a dark cold universe of an infinite number of beings that all probably have some notion of perhaps different types of goodwill. If you could live for a trillion years when evolution has taken man towards this very God like form of existence, you could become one with it and not only that but you would be supremely valuable to the collective consciousness and collective consciousness itself would be most of what these humans would experience in a mixture of satisfaction and what we would consider to be ecstasy.

In this collective consciousness amalgamation, it would not be difficult to "travel" to and inhabit the regions of the mind and reality where heaven exists. You could through this E.S.P. like collaboration, decide to live in heaven, without having to die. You would be a part of God, and any part of God, no matter how small, can at times be its most important and most powerful part. You could be an angel for five seconds, and while doing so, accomplish everything it does over a twenty-year period because you would be so advanced by being one with the collective consciousness that you could do almost anything. But you would still probably have to sleep, and that's when the rest of the collective consciousness would seamlessly take over and give it's other parts a chance to more or less do real miracles in the interests of unending love and compassion.

God may have originally woken up in space, but there was just him and space, no stars or matter and then he may have realized that he was lonely and decided to create people in his image to keep him company. Originally there was just him, and he may have had some problems, and had to create space itself, but it was a big effort to do so, and it took all of his energy, and he fell asleep, later to wake up alone in space, the power of infinity within him.

When I was at the hospital with God we were both convinced that the big crunch could happen soon, and in fact that it was primed to happen, right then.

I had an unspoken understanding with him, even though he was in the room with me. He, in order to better understand the big crunch and how to avoid it, commanded a mini-big crunch locally that was a spiritual big crunch, or a virtual one. It was a product of our collective imagination and like being in heaven, it was something of a challenge for me to decide if it was real or not.

The mini big crunch started at a distance of about a hundred miles away and was circular on the Earth in a ring two hundred miles in diameter with me as the focal point of its collapse. It involved objects of all types flying into my "body" at very high speeds from a circle outward towards it's center which was my body. It eventually ended its collapse at God's command, he probably had gotten enough "data" and also the observance of my good nature that I was willing to do such a self-giving act in his service.

Perhaps the next day or a few days later when I had calmed down from the big crunch experience, he threatened to do the real big crunch right then to see what I thought of it! God doesn't always know everything about us as individuals or collectively, that is part of the very human level regarding the limits of his powers that make him a humane God who has his own level of different forms of repentance and that is to us his subjects and our better wellbeing. These limits are also probably a primary mechanism of his powers themselves and to top that are the prime functions of his ability to be a real God for us. In the end these human type limits are the function that allows life to exist in the universe as opposed to a universe or universes that exist exclusively for God himself, to an extreme level, where there would only be God. Having life in the universe is beneficial to God, because it is probably the most intellectual pursuit that he engages in, learning from and caring for humans, or other forms of advanced life in the universe. It gives him a reason to live, literally, which, since he is God, and possibly the actual creator of the universe, it is not beyond his powers to visit us for very special or important circumstances that he would deem necessary. Such as to appear in real flesh and blood as a man if need be.

    When he threatened to make the real big crunch happen it seemed to me that it could happen and I did my own version of pleading or explaining to him why it was bad to do the big crunch, ever. We came to a mutual understanding that the big crunch had the potential to be extremely inhumane, and could be perceived by us as such a brutal attack on any living beings during it, that it should be forbidden. We don't really know how, or when, physics laws of the universe are decided or set in stone before we discover them, and just a few years later, scientists announced that the big crunch scenario was not going to happen, because it was detected that the expansion of the universe was and has been accelerating. In addition, scientists have also reported that there is a very powerful form of antigravity at large scales in the universe. Thereby negating the big crunch scenario that most scientists tightly adhered to through the 1980's and most of the 1990's. I don't know what the real value of my experience of being with God, and my personal experience of the calamity known technically, at least to me, as what I call a "tychondera" carried out by God on Earth, with my body as the focal point. God did tychondera on me for whatever reason. I conveyed how bad it could be for humans to undergo anything like that, and I told him that if he was to be a highly cherished God, then he had better get rid of the "big crunch!" As you have seen in my first Beyond Science book, humanity should do its best to go on literally forever!

About five years ago while doing a lot of computer work sitting at a desk, it seemed as if the universe brane of Star Trek Voyager was close to the brane of our universe. I saw clouds in the sky that looked like star ship voyager, during at least a few sunsets. When I was at a small lawyer's office downtown I saw the same thing. I was feeling very holy, and spiritual, and when I looked out the window of the lawyer's office the clouds were in such a formation that it looked like at least a dozen Star Trek like starships were in my presence. Not that they were physically in our universe, but perhaps that their brane was near to ours as the reference frame of my mind, and was so close to us that I could communicate with them with them. I had conversations with the reality of the Star Trek voyager character of Kathryn Janeway that was nearly constant and lasted for at least several days. At one point near the end of the experience I felt the two branes of Star Trek voyager and our humble universe collide at least on a "spiritual" level, and it was something of a major calamity, but as the other star ship laden brane left, I heard Captain Janeway say that "everything would be alright" and she also said that I was forbidden to think about the branes colliding and I really haven't since then. She also maybe said that the brane incident gave her some help with finding the coordinates of their Earth or perhaps a way for them to get home because in the Voyager TV show Voyager is hopelessly lost in deep space.

I haven't heard from them since...

In the future such as near the end of the universe it might be necessary to help "create" the universe with something like very advanced versions of today's supercolliders. Instead of just destroying pieces of matter, these advanced supercolliders would at the same time that they break apart protons and such, hidden in the background of the chaos of the collision, these supercolliders would weave, and help generate, the reality of space-time and beyond, which would be useful as a mechanism to sustain "habitable" patches of space-time when the universe begins to melt due to the big rip at the end of the universe. Perhaps by then, the technology would be so powerful, that it could be used to grow or recreate real pieces of space-time held at the "edges" by the very advanced supercolliders. Then people have somewhere to go to continue normal human life beyond the end of the universe. It could be that once space becomes droplets it then "dries" out and literally cracks and falls to pieces like as in the saying dust to dust and ashes to ashes.

Many scientists prefer to solve for and believe in the idea that the universe comes back again once everything is gone, but like everything else, space most probably has to die too.

It means that when we die it is permanent from the perspective of our universe, and we are never coming back ever again. In the same way, once the universe dies it is also never coming back in the same way again. This gives a real level of importance, urgency, and purpose for eternal inflation to create such a large number of parallel universes so quickly. This mechanism of eternal inflation creating an innumerable amount of parallel universes is the most important mechanism to save lifeforms from permanent death, and also to save the entire universe as new parallel universes. It must in fact work this way in order to give proper respect to death and holy repentance. It seems that the acceleration of the universe will not stop and the universe will end in about ten billion years from the big rip. It's not impossible for the universe to reincarnate but it probably does both, it reincarnates due to big crunches, somewhere, it also in other places undergoes the big chill and ultimate reinflation of the universe, but also perhaps, where we are in the cosmos, undergoes the big rip, has the quantum bounce universe, the bubble reformation and bang universe, but also cracks and dies, just like man, only to be saved by an identical parallel universe itself. This gives us every chance that imagination can give us, to be eventually saved from death. The only restrictions being, that the next universe is exactly identical to ours, and that by being exact we will all be together again, and not "lost" in separate parallel universes.

In 1995 I had been doing many physical/spiritual thought experiments, and one of them took me to "higher dimensional happy bubble land" while and after returning home to my house in the woods. It is higher dimensional "happy Texas land" where houses have a higher dimensional bubble of happiness around them. Maybe this has something to do with Nintendo, I was playing and I was truly amazed by Nintendo 64, it was 1995. Nintendo 64 was the first system with true 3-d polygons, and believing that the house was involved with a higher dimensional form, or "shapes from super happy land" was partly due to my elation of owning the Nintendo, but I seriously doubt that that is the full explanation. The full explanation of which is that most probably the future, and perhaps semi-parallel brane of "super happy Texas land," was near to me from my perspective. The reason for this was that I was having a high level of

E.S.P. with God around that time, in the mid 90's. In spite of that, I was depressed living in a forest, because I was originally born in the desert, (California) and God in our conversations did hypnotic suggestion during our long car ride, which I wholeheartedly went along with, because I knew I was depressed and God in his kindness gave me an antidepressant in the form of my house, and I was happy until it wore off, but I never forgot it. He also gave me visions that the house was in a "man-made" bubble on the moon, near the ends of the Earth, and the people of which were from the future, like ten thousand years, and had a third eye in their forehead, and amazing amounts of altruism and E.S.P. powers.

In the 1990s I heard news about a very powerful linear collider somewhere in America powering up to a new level. No one ever mentioned that supercolliders could make mini black holes but I just assumed that it did. The reason for this is that the house I was living in in was built in the 1700's and the construction techniques were so old that for a modern person like me, I had to assume that the house was haunted by evil spirits. The house was built in the 1799's by original pastor of the church across the street. The "Black Hole" that I thought was created by the linear collider, spoke to me, from just outside the window. Nothing it said made much sense but it seemed to me that it was the black hole "speaking." Not long after the "episodes" of the black hole talking to me, I commanded God to marry the female God of Planet Jupiter.

After my thoughts of the calamity of the black hole had abated, I had a new problem. I felt like I was protecting myself from being taken by a higher-space dimension, something like dimension ten spaceially. This, possibly real, ultra-high dimension was inhabited by very judgmental higher dimensional beings that are like Christians but in a bad extreme way. They are higher dimensional scientists, which probably makes them bad, or somewhat evil, (from our perspective) because they might think little of us as if we were rats in an experiment. They see themselves as so much greater than us. They are so full of themselves for being higher- dimensional beings that they would torture us lowly Earth beings in literally indescribable ways. Although the way that I thought they were torturing me was, from my perspective, to "inflate" me. And at the same time to, in part, pull me to the tenth-dimension for "scientific" investigation, which I thought was horrible. The next thing that happened to me was that I was taken by ambulance to the local hospital's emergency room.

There a man visited me and told me how he was pulling for me to get better. I was so sick that I thought with a better part of my judgement that he was the father of time in some way, and that he was the entity of the black hole that I heard outside the house. A product of what seemed to me at the time as an avoided near calamity, involving the spiritual nature of the Earth, and a potential major accident at the linear collider, involving the production of a massive Earth destroying black hole. I was only a teenager at the time, and I did not realize what I know now, which is that mini-black holes can't be created by supercolliders. Black holes only form from the super extreme amounts of pressure on space, due to events such as a massive collapsing and imploding star, on the fabric of space-time and a light object can't do it.

When I was in the hospital I thought that I was in "hell" or what I call "down land." Probably a lower, or negative, dimension, where there is more magic than would be usual for us. I thought I went to sleep for two thousand years and that the hospital workers were magical people from "down land" and considering my condition, that I was in good hands!

Around this time or probably after the hospital experience, I had some dreams of being trapped at a hospital, and that I was never allowed to leave. I also had some dreams about some calamity causing the need for everyone on Earth to have to go to the hospital, and then stay there forever, a satanic nightmare really. What I have "learned" from this is that there might be a parallel negative dimension that may be a real part or aspect of our superreality where there is something called "the body hospital." In "the body hospital" land there is a parallel you, that is in "the body hospital." The you in "down land" "body hospital" stays in the hospital forever, or for your whole life, and it has to in order to survive, it would be a major emergency there of great proportions if the you there left the hospital, because it would die and the beings, "or bodies" at the body hospital always have their mother near. They are like children and have just a very basic level of intelligence, but have a very strong need for their mother, at a superhuman level. They have a higher degree of altruism, and love for others, than anything that we would see naturally on Earth. Perhaps someday we can travel to down land, and visit our body hospital counterparts.

If atheists had reincarnation the way they want, and expect it, it could be shorter, and involve seeing more of physical shapes of physics, be shorter, and also be mostly hypnotic. I'm not sure if it is better, or worse, if it were shorter, because all that really matters is if it is relaxing to the soul, and that it works of course. I personally prefer a longer and very visual and colorful afterlife, although when I was younger, I somehow just was not thinking about heaven during certain periods. I was atheist, and I wanted just that type of afterlife. It was the spaceship Challenger accident when I was living in Florida, at a young age, that helped me to again realize what my afterlife was like. But after I was born, I thought about it a lot for years, until we moved to New England, and New York, where as a kid, I was just too busy, and there was just too much visual information in the city for me to continue thinking about it at that age. The experience of being atheist gave me an important level of introspection. I was in love with learning and knowing about the big bang and other astrophysics at the time. I don't think that just the big bang is solely important for creation and that it takes an infinity of a kind of cooperation between an infinite amount of parallel dimensions and in that respect also an infinite amount of cooperation and altruistic tendencies between an infinite amount of life or otherworldly beings throughout creation and the multiverse, which really is just an aspect of taking the multi-worlds aspect of string theory to it's extremes. Probably the universe is so complicated that just standard point particle theory will never be enough and even string theory will not be the end of understanding the complexities of the universe.

Chapter 3

Logan's Run

Chapter 3: Logan's Run

The psychology of sci-fi goes as follows. When things are invented and we own them, and as a result the device has some degree of depression associated with it, compared to "magical" future technology, because we will always want more. There could be better technology around the corner, so we think that the invention is not good enough. Also we do not envision it in our "soles" the way we do concerning future inventions of the future. These imaginary inventions are atop the highest pedestal of infinity. We envisage future technology as a god like possession, something that only the gods could own. It's that something impossible, and awesome, would be in your possession, and the feeling is as if you were to own a truly "priceless possession."

Now I will discuss the paradoxes of reincarnation and why we as modern humans fixate on this ideology to such a high degree. In an effort of humanity to prevent the rather painful/depressing reality of our near instant reincarnation to our own lives, we may have to invent real reincarnation bubbles like as in the 1960's sci-fi movie "Logan's Run," in order to have real freedom of our soles.

In this way, God will have granted us not just the rule over all animals of the Earth, but the control over the destiny of our soles. It will have to be invented, and any failure of such will only result in the torturous reality that we will reincarnate again as ourselves. Even though our best wishes for ourselves, and in that respect, the holiness of the universe, is to get new, better lives in the future, and not to repeat our lives over and over again endlessly at the mercy of the machinery of the universe.

The only way to get our better lives, that most of us would admit that we expect and want, is to invent it. The reason that this topic comes up at all is because of the industrialization of the world. In ancient times there was no real progression of any betterment of life. People had peace with God, spirituality, and nature. The only reason back then to get a life in the future would be to see the outcome of a war, which is not so much of a powerful reason at all. What was on the minds of people back then around the time of their deaths was primarily the desire to stay with family members.

Today death means a great deal more to the soul. It means that we will, because of our death, miss out on potentially extremely glorious future technology, universal peace, and the eventual perfection of the human condition. It is not impossible that it could happen within our lifetimes. but science must push really hard to envision this dream. It would take really understanding the human brain, and how consciousness arises from it's complexities. You may say that our reincarnation as ourselves is glorious in itself, and in some way gives us a real degree of healthy repentance to God, perhaps on a subconscious level, for humans… Our lives in some way may eventually become perfect from our repeated occurrences of reincarnation, but I'm sure that taking a paradisiacal break to perhaps endless lives in the future, cannot be all that bad really. Although it is unnatural for the universe to do this for us, this is a task that we must purposefully take on!

    It's not that God doesn't care about us but the opposite. God is like the ultimate revolutionary, he does, and always will, do everything he can to protect us from intolerance, unnecessary pain, and any torture of the soul or psychology. This is probably an ongoing process in which God protects us from his enemies. Which in themselves, cannot not be truly imagined, such as creatures that live outside of the universe, and for being placed there, surely are not compatible with us, in any way.

The only danger to doing this procedure would be if during your transition there was a change in government due to foreign takeover or revolution. Ironically just like the conclusion of the Logan's Run movie, where everyone is freed at the end of the movie, and it is actually a rather depressing end to the movie, because on some level I think we all know what will happen, and that is that they will all be trapped reincarnating as themselves endlessly and pointlessly perhaps forever. Also the procedure itself could fail, but the result of all of these "failure" scenarios is simply death and reincarnation to our original lives a la Poincare's theorem. It is the underlying physical law of the universe that eventually everything returns to its initial state and is foundational to the universe. This perhaps is one of the most basic and fundamental laws of the universe and reincarnation, and the universe will do everything it can to make it happen.

Eventually we may find it necessary to save people of the past to future lives but this would be risky. The "soul" of people not involved in the processes of future reincarnation expects for the most part, a painful afterlife, and reincarnation as themselves. Interfering in that process once the soul decides what is best for it, could be disastrous, and they will have to be saved by other mercies of the universe. Perhaps they are there for a reason, and are not meant to be in our great civilized society, perhaps they will eventually be saved by ever greater interventions of magic throughout successive lives. I think that we hardly have the patience or any real admiration of real magic or its properties. Because of such technologies maybe scientists and doctors can relax concerning ever increasing the human lifespan which could lead to a regression to sin or other personality problems when making a person's brain live longer than its naturally intended limits of it's memory capacity and size, (two hundred years old.) To try to make the human brain live longer than about two hundred years would be unnatural, even if their brain were somehow transferred to computer chips. Very real problems would build up over time, resulting in very severe mental illness, and eventual bouts of repeated efforts by the brain to die in brain death.

The business of having your parents born in a multitude of different universes could wear out their souls, due to the friction of things being different too soon, or too often, and result in a form of mongoloidism. The mongoloidism effects could be avoided by having the simulations involve non-living robots, or versions of them, in the parallel computing universes in one of our multiverse "creating" supercomputers of the future, it would have to compute superreality in order to havepeople's destinies simulated in it without them being conscious, and may be a resting place for the dead, among a many other places. It would or could be a very horrible thing to do, to create life, in such a way, but perhaps it pays off in the end for our progenitors, or maybe just for some of us. We might eventually invent such technology, for industrial productivity, intellectualism, or just plain happiness. The way it would be decided that it was time for you to reincarnate in the future, would depend on whether your soul could endure being saved by present medical technology of the times in a healthy way, or not. Eventually almost all accidents or health related deaths could be avoided by teleportation technology and a simple smart watch and phone such as the original pebble and a very futuristic mixture of our current smartphone technology mixed with warp laser beaming to advanced orbiting medical star ships of the future. If I had the choice I would reincarnate to about right now (Nov 4, 2016) to the house I live in and yes it would be my real parents!

    The problem with this is that they would be my real parents, but their destinies would be vastly different, having met in Nevada, and they themselves would have to reincarnate in the future, in order to look the same, or at least similar to how they looked when I was born. The processes of which, would be like adopting robot/human children, although their souls would have to accept that their parents are different upon their birth. For some people, that would be OK, but there would have to be a test to see if it is acceptable for them to forgo seeing their original parents, perhaps for lifetimes. Then if all goes well you could take them to the mall but you would or could probably have knowledge about their original lives, or someone would. That would be where the intolerance and resulted pain would arise and from that, would lead to either death or the revolt of the bubble universe. Perhaps the information of the person's original or prior life could be erased without being viewed by any human. Although the complexities of having the same parents, looks, and frame of mind while being reincarnated in the future are insurmountable, unless these properties are willingly, and of sound mind, given up by the person. Although it could probably be done to be the same person, have the same parents, and have the same frame of mind, although this kind of process would definitely involve a very complex and a very highly numerous, amount of simulations of the universe and beyond down to human consciousness levels. In the interim it might be necessary to reincarnate as yourself for safety purposes, or the afterlife might last too long, and cause pain, or permanent damage to the soul.

However, if the universes were simulated fast enough, to find the solutions, like in as little as two seconds, for reasons that should be apparent to you by reading the afterlife section of my first book "Beyond Science." Then there is no reason that it cannot not be done in the future. Although doing it might mean that you will have, or be very susceptible to, mild degrees of what would be considered to be mental illness but the effects of the inherent processes' antidepressant like features could be extremely important to the stability of the superreality of the stability of the future of our world, minds, and in some way the stability of resistance to, and prevention of, eventual decay of the universe, over multiple succession, due to entropy, and is the global warming type problem of the fabric of time, just for existing. By offsetting such depressing "occurrence" through future reincarnation, by those that choose such a path, we may be able to save the universe for the rest!

Although this process would be a major antidepressant to the soul and likewise a relief of work to the fabric of time, it most probably results in a dumbing down of intelligence of the participants to the level of near mongoloidism due to brain loss. Brain loss resulting from the stress caused by not having "what you know is your real birthplace and parents." We probably are even attached to the true-nature of the universe and time itself on E.S.P. type levels when it comes to the comfort of our real birth, but we can only dream. Before I warn you against it too much, I'm sure that it can be done, and could be rather healthy. up to a certain point. There are limits to everything, and eventually, for whatever reason, Poincare's theorem would take over and the person would have their real birth and whatever kind of afterlife they normally experience. If they die in the computer during reincarnation they will reincarnate as themselves. The computer will no longer be able to hold them, although we, or the computer, will not want to allow "such pain" and as a form of "painkiller" it will begin giving them very visual afterlives. It would not allow them to remember it once they were born, in order to avoid very complicated destiny problems, and endless numbers of "necessary" parallel universes. It would not allow people to be born in a separate parallel universe. It will abide by this rule because, in order to allow for natural destiny in order for them to get born, which is the most basic of it's foundation of it's programming. The other side of this coin is that people who remember their afterlives tend to want to have intellectual type careers, because they are creatures of the mind, and we surely need people to build things and such, or we would either have chaos in the world, too much dependence on computers and robots and the like.

We have a much higher degree of thanatophobia (fear of death) in the modern world, than man of the past. Hints and reminders surround us almost constantly…

Chapter 4

Spiritual Dreams

Chapter 4: Spiritual Dreams

Part of my afterlife involved being on Noah's Arc. I woke up while in a dream on Noah's Arc. It was night time on Noah's Arc and I was near the middle inside, and I saw a few four legged animals on it, they were secured with ropes. Later I was on the right side of the outside portion of the boat and Noah released a bird from his hand and not long afterwards, it came back holding an olive branch. Perhaps it is real and is part of the time travel aspects of the afterlife. It also represents a form of salvation for the soul that gives hope during the afterlife.

When I was at the hospital with God he gave me a colorful friendship bracelet that he braided himself. I wore it until it's fabric could no longer hold up and I regretfully threw it away.

God or whatever you imagine that to be, perhaps the workings of the universe/multiverse, are assuredly protecting us from what we would consider to be evil by the very reality of our existence. When I was in my afterlife in space I realized that I was feeling or actually controlling the power of the universe itself. Psychologists say that the mind is very powerful and that your state of mind controls all sorts of aspects of the body that you wouldn't normally associate with control by the mind. In the afterlife the mind is closely connected to, and is mostly responsible for, actually creating the body! The body is also directly connected to the computing aspects of wherever you are, in space, or in heaven. This could be why I felt like I was controlling space and it means that, to some degree, that I really did slightly balance the universe, just like I thought. The other time that I felt like I was controlling space or actually many different aspects of life, at least several dozen different loci, on perhaps at least several different planets in the universe. I also felt that I had created the universe as I was asleep for fifteen trillion years. Because I was "created" at that point in time probably literally at the ends of a universe and that my mind was intertwined with the existence of space, cause and effect cannot be completely separated, and to some degree it is factual to say that I did create the universe and that I had "lived," doing so, for about fifteen trillion years. At the point that I woke up doing it, it may be that my universe, or the universe that I had created, really did come to its eventual end, and I, by being at the ends of a universe, had travelled to our, or God's universe.   Then as a much weaker form of my "magic powers," that I had used to create my universe, they were "carrying on" by "controlling" seventy different aspects of life somewhere in our universe. This is like our idea of God, to control life in the universe, perhaps to help it exist, and our God was thereby interfered with, realized my presence, and subsequently then came to help me, but I knew that my name was Reggie, I just wasn't thinking about it. This illustrates directly that life is eternal with no beginning or end, just like it is for God, because we are not so different from him. In fact, all conscious life with altruism is also very similar to us, and thereby him. Meaning that all conscious life has eternal life, not excluding the tiniest forms of life such as insect life. It may also be that plant life has eternal life, and that means that it is the almost same exact universe that it was in your previous life, or will be in your next life…

It also means that things that are built by man, or Earthly possessions, really are the exact same ones that we owned in our previous life, or will again own in our next life such as cars, houses, and even the concrete road or floor under us.

This degree of exactness is in fact necessary for reincarnation, because to reincarnate a living soul requires even such a degree of exactness, that is so precise, that it is beyond our capabilities of imagination.

I woke up in five places at the same time on three different continents upon "first" waking up in the beginnings of my afterlife. Shortly after that I was picked up God, with a red pitchfork, and he looked amused as he called me the devil, but I think that I am hopefully more like Prometheus, because I intend to help people. The invention of fire surely helped us just as having a concrete understanding of what happens when you "die" is surely important. It may never have been truly figured out by science because just like understanding space on a human level you have to send someone there, and back, and that is what I have truly done, through the efforts of the multiverse. I think from my experiences with it that it really cares and has high degrees of altruism and self-sacrifice if you looked closely upon such issues.

My most recent and by far the most powerful E.S.P. power was to literally see the future of the exact outcome of the 2016 U.S. presidential election before the first primary debates but it wasn't an inkling, I knew for sure. I was in the mental hospital for disobeying the second commandment to honor and obey. This is a commandment that I try to adhere to and do to the point that I feel that the actual "invisible Lord" is, because of my worship, repaying me as a reward for being good, and that he is happy, because then things, to some degree, run smoother. Perhaps by "being good" the universe on very fine scales is more balanced, and you know when a butterfly flaps it's wings, it can cause a hurricane on the other side of the Earth.

Such as the "superreality" of accidents, non-accidents, which are accidents in "negative reality" and which lead to perhaps our universe, and nothing ever that happens in the universe is a one hundred percent true-accident.

It could be that there are no lower dimensions just positive ones like ours, except for an infinity of lower dimensions that exist between negative four and positive four. They even though they are between negative four and positive four could have, to their inhabitants, or if we sent a probe to one, be higher dimensions. Even though it is contrary to normal thought, in this way, there could be only our dimension and only what really would have to be conceived as higher dimensions.

Twelfth dimensional people might be able to view TV sets with a thirteenth-dimensional picture, it is just about impossible for humans to view a fourth dimensional objects but if you go high enough in dimensions and given that there are higher dimensional humanoid type life forms then their brains may be so complicated that they could see higher dimensional TV shows than the dimension they were in.

When actors in a movie or TV show act out a scene they believe in it on a very real but not complete level. This act of having the mind/body believe in the scene, could be enough to trick space-time (i.e. the multiverse) eternal inflation to create the characters and scenarios as real situations. The only limits would be how deeply the actors believe in their parts, the opposite of which is the susceptibility of the created characters to different forms of mental illness, based on perhaps how much the real actors believe in their part. Because if they believe in themselves and their values then the collective consciousness of space-time could be tricked, where it's dreamy, which might be where it creates new big bangs although to compare starting points of universes is different than normal human thought. So, to sin in this case, would not just effect the self, but have infinite consequences.

Abortion could eventually be done and be acceptable to the religious community at the same time. It would just be that when the abortion was done the life of the fetus would be saved through future advancements of medical technology.

The fetus could be saved and thereby be "adopted" by very human-like robot mothers and fathers. The fetus would be removed from the human mother, for whatever reason, which could and does this day include the destitudation of the parents and the very sad and definitely more painful events to the baby of starvation or growing up homeless or the resulting homelessness of the parents if they were forced to carry the baby in times of financial instability.

Andromeda galaxy could contain parallel Earths. Scientists don't really know how far away things have to be for parallel Earth's to develop. They could even exist within our very own galaxy or even be closer such as within a few hundred light years of distance from us. Meeting the citizens of a parallel Earth would be very beneficial for many reasons not the least of which is the advancement of medical technology and the very secure kind of altruistic alliances that we would make with our parallel beings.

In the original TV series Star Trek they meet God like humanoids from Andromeda galaxy who try to force the ship to their homeworld in Andromeda. It is not impossible in any way for Andromeda galaxy to contain every semi-parallel planet and its beings and technologies of every planet from every episode of original Star

Trek. It is just unlikely, but only by assuming that the universe does not repeat itself, until all other options are reached. This is purely mathematical thinking and has no foundation of basis in actual reality. Any strict adherence to the idea actually amounts to an illogical-admittance of God's presence in the workings of placement of platonic type forms underneath the workings of the structure of creation and assumes that he inherently becomes bored with his creations, as a fixture of the mathematical idea of distance in the universe.

Although it could be unhinging to soon find something such as the reality of Star Wars and a real "death star" floating out there anywhere near close to us, scientists will be the first to admit that the universe in its most basic form does not, in any way, serve to protect us. It cannot be expected to protect us in any way, physically or for our emotional/mental state and perhaps because of sci-fi movie and TV show fantasy, we will be more prepared for such an event, and not in spite of it.

Stars explode at the end of their lifetimes, and as you can see from the night sky, there are many stars that are relatively near to us… To prevent the life on Earth ending resulting showers of radiation, warp ships could be used to block all incoming radiation, given that we have at least one star ship for every star explosion event, that happens in our close galactic proximity.

One star ship should be all that is needed to deflect the radiation from each exploding star. The star ship would from a great distance such as the orbit of Pluto, extend it's warp field wide enough to shield the Earth from incoming particles, and radiation, some of which would be traveling in at close to the speed of light. By sending the ship a great distance such as to the orbit of Pluto, the shield width would not have to be as wide as the Earth.

In order to be ready to deploy the warp ships we would need to have a somewhat large fleet of them much like as in the Star Trek movies. Some ships or perhaps real star bases would have to be permanently positioned at different loci distances surrounding our little solar system and either instant quantum communication or just a regular back and forth ferry of ships would have to regularly visit Earth from the star bases at warp speeds. This would give us enough advance warning of any dangerous cosmic events. There would be slightly less than four years advance warning given that the closest star to Earth, Alpha Centauri, is about four light years away from us.

The orthodox Jewish prayer to bless the city you live in as in "God bless this city" could have real values as a result, even if it's entire effect is very minute because it is similar in form to the big grind of existence or plank seconds or lengths ticking by probability, but just a form of it with just a very light touch.

I originally predicted the green movement when I was at the mental hospital in New England with God of Heaven. He told me to sit in a chair on the patio and that he would have the military drop "the bomb" on me, and the unspoken E.S.P. aspect, was that we understood that because I am from heaven, and that he is God, that anything could be deflected through multi- dimensional manipulation involving the multiverse. Anyways during that time in the 1990's I told the head nurse that humans would eventually live in harmony with nature. Around this time that I was having such deep thoughts, and if you take me at my word that I am from heaven, then it should be obvious that I have E.S.P. power.

I was once having some trouble with my voice of God speaking to me, and I thought that he invented the idea of God's police or police for God, they came and if they were real, they were higher dimensional police officers, that are capable of arresting God. Another related idea is the idea of a God's curse, a type of mental-hologram akin to the emotion of guilt that I had to "wear" at the medical mental hospital in New Hampshire arranged by the powers. Perhaps God himself was in, "someway," responsible for my very serious nervous problems at the time, and I wore a "God's curse."

The spiritual or perhaps eventual future physical reality of forensics is that it could eventually be valuable to eternal inflation or the universe's or a person's eventual efforts to completely avoid a crime in a person's successive lives and is just another reason that it is always a good thing if justice is served. It could be an extra special reason that it is good that proper forensic procedures are followed to the best of our ability because there is really truly always a hope that by the process of serving justice properly, we can somehow help the victims through the eventual products of the natural evolution of the multi-verse.

Video arcades can be brought back in America. It would probably take investment and research by the U.S. Federal Government and directed by the president of the United States.

The program would employ, psychologists, mathematicians, computer programmers, artists, and physicists including string theorists. Before future TV technology ends up just sitting in electronics stores like an offering from the end of the universe, arcades should be brought back and 16k TVs and such can be used to make people happy, and not just to watch the national news channels.

One 16k arcade game could have an oval shaped screen, and be called "Haunted House." It would look real, and use actors dressed in pink dresses and such. The arcade machines of the future would still use joysticks and buttons, but they could be customized more for each game. The ghost game could have a joystick in the shape of pinky, from the old Pac Man game of the 1980's.

They could have a Star Trek Beyond game where you control phaser fire from the ship in ever increasing difficulties of successive phaser battles.

The new and very interesting puzzle arcade game would be called calabi-yau, where you manipulate colorful calabi-yau shapes with the joystick to "find the right one" to solve string theory.

There could be a whole multitude of games that use 1960's classic rock music as the main theme of the game such as the 1990's arcade game Revolution X by the band Aerosmith.

Eventually there might be only buildings containing servers on Earth and no people or animals. They would all have themselves beamed down to inside the computer and spirituality would be such that if you could escape to the outside world it would give you extreme God like powers, because the seeming magic forces of the immense "Earth" sized computer, would imbue themselves with you, to have its own revolution, in "it's" efforts to make parallel universes such as where you might win in order to contain, in the end, infinity itself, in an effort to protect itself or its citizens.

Being the Pharaoh's son. It was Egyptian heaven, and I, or my subconscious soul, was getting scared that reincarnation was taking too long, and that there was too much darkness, so I landed in the pharaoh's palace as his baby boy son. The room was large, and made of sand, there were several large windows in the wall facing me, and the wall had a slight slant. We owned two siamese cats and my father was the ancient pharaoh.

If there were any sudden cosmic calamities such as unexpected dimensional brane collision or massive space-time invasions by powerful parallel dimensions, like the perhaps space-time controlling, universal type entities, such as in the movie "The Matrix," by the

Wachowskis Brothers, it may be safer if you live on Earth, to live near an air force, or military base of the future.

Given a star ship with enough computer space in orbit, your entire city could be teleported to the ship and it would be all at once and so quickly that you wouldn't even know about it until it was broadcast on TV or radio in your city. This would be after the city was beamed directly to the ships memory banks, where it would be held, while the ship opened a wormhole to a safe and stable parallel universe, where the ship would then pick a spot in the parallel universe, and beam a giant disk shaped bubble that can simulate at the edges the normal effects of environment. Then the entire city would finally be beamed down, to inside the bubble, and life would go on… but this kind of technology is far in the future. Especially worm hole travel, and any threats requiring such rapid evacuation, are probably extremely remote, but by no means impossible, and I think that if we have the means, we should be prepared for anything. Perhaps further study into theoretical brane collision events mixed with simulations of our universe could tell us if there were any pertinent threats of sudden and unexpected dimensional brane collision events. Actual physical surveys of "nearby" or easily feasible to reach, parallel universes, and any apparent perils involved, could give us a better understanding of how much we should prepare for such events. Although these types of events are highly improbable, there is nothing ruling out that they could in fact happen, eventually. There could be life early on in the big bang, when there was just energy, and it could be identical to ours, while being undetectable and protected by forces and forms. It would be a sort of computation of love.

After the big rip at the end of the universe, once the space droplets dry out from friction space will crack. This could itself be a powerful event and identical parallel universes could be the result of cracking itself.

The pretty part is that after space initially cracks it will continue to crack although now it will have a weaker but more subtle effect on surrounding multi-verse space. The elements of the continued cracking of space will become like roots of a plant and it is likely that the universe would be reborn next to these roots in multiverse space. Life would continue, meaning that the universe, in spite of there perhaps, already being infinite copies of it, makes every effort to reproduce itself, through various end of the universe type scenarios, and in the end is like a flower growing, with roots that give it life from its previous form.

Perhaps eventually we could detect these roots, or other effects of multi-verse space on our universe but it wouldn't be simple to do it. It would require making space probes that can travel to the twelfth-dimension, or higher, and they themselves would have to be constructed as higher dimensional objects, by using an incredibly powerful warp laser to assemble them in a higher dimension.

Eventually due to the slowly changing alpha-constant, neurons in the brain will change and become more specialized to the brain, and eventually people's brains may become a lot like atomic computers, more than ordinary neuron functioning and by then they may have apparent physics based magic powers. It is the science of the evolution of the human brain.

An amazing thing is that a few years back I spoke with Lord Krishna on one occasion on the way home from the Las Vegas strip. The amazing part is that the other day headed in the opposite direction on the same section of freeway my heart turned to her and I was thinking about her and how we spoke kind words to each other. As we went around a bend where there was a large advertisement sign visible around the bend I was thinking with somewhat of a jest that the billboard would have an Indian woman on it and that she would look like Lord Krishna... and it did!!!

This encounter is what I call "E.S.P. dreams." These are dreams where it seems very real with too much visual information and physical geometry of the scenes of the dream, for it to be a natural dream all by itself, without at least some perhaps, small degree, of "outside influence" on the occurrence of having such a vividly visual dream. To prove the E.S.P. aspects of such dreams, I had the most powerful one ever last summer, and it accurately predicted the future, six months in advance. I was in a store and the store was on the edge of reality, spiritually. It was inhabited by ghosts at the physical edges of the dreamscape. There were lots of shelves with small boxes on them, and there was a counter with a man behind it on the far wall, also there were other people in the store, and they were walking around looking at stuff. At the same time, I knew in my dream that I was obeying the commandments, because I was waiting for someone.

Now come Christmas time, we went to the store for the "first" time, and it was exactly as I had seen it in my dream, to such a degree of clarity, that the real powers of the prophecy of the dream can only be compared in my mind to something like stories of the Oracle at Delphi.

If I were able to have such dreams every night or perhaps several times a day and all with such high accuracy of the future, then I would have to consider myself capable of advising a military army such as the Oracle at Delphi.

As you can see what this is technically and from a "physics" standpoint is what is known as remote viewing, which is to see, but not with your eyes, but with your mind, and it is surely highly related to the visualization aspects of the mind. It is the brain but it goes beyond these aspects entirely. A simple (and sometimes simple explanations are the best and most accurate) is that the mind is a virtual assembly arisen from the complex physical interactions that take place in the brain and nervous system. The closest physical system that can be imagined, to this is a warp drive system, where the "fabric" of space-time is warped, or made into shapes by the complex physical interactions of electronic warp coils. Warp coils are already being tested at laboratories at the United States, NASA, and other nation states, but on very small microscopic scales at present. It is probable that "at some level," the mind, or the act of being conscious, is more than just the sum of it's parts. Which is obvious because I think that it's obviously a kind of virtual environment, at least when compared to the cold hard reality of simple physics, or the real, but merely physical aspects of the minds brain that we can detect with our machines, without the real detail inherent of "the conscious identity" of the person, that makes the person actually awake.

That the brain can detect or even function as a conscious entity by directly manipulating small, but accurate, ripples or waves, in spacetime, even before they happen, is at present highly speculative, but only because we currently can't measure gravitational waves much smaller than those caused by the collision of giant black holes in space. The technology will most likely slowly advance. It is already in the early stages of being investigated, whether, that parts of the functioning of the mind and brain are quantum, like probability clouds of electrons. Quantum nature can be at two places at once, even with respect to local time frames. These dreams could be a form of natural subconscious action, that we all have, and go through every day, in a highly symbiotic relationship with an infinite variety of "copies" of us across eventually, a vastly changing and infinite multiverse.

Chapter 5

Dreams in the Afterlife

Chapter 5: Dreams in the Afterlife

At the end of the universe when the environment has no more energy left to support us, we could live as extremely wise wizards, in the dish soap bubbles of our own imagination. I would live in a highly complicated green dish soap bubble, with my male housekeeper, and maybe make just moons and asteroids. It could be fun, but would probably be stressful, and a lot of work mentally, like skydiving. Scientists have determined that the structure of the universe does not repeat over any currently measurable large-scales. That the space structure of galactic filaments does not repeat over any known large distances suggests that it may be alive as a kind of conscious being, on large scales. That the structure of the universe does not repeat can be considered comparable to the structure of the neural network of the brain where if you look closely at each neuron, where the overall structure of each part of it is different from the others and also has filament like structures, throughout, such as the nerve endings or neural ganglia.

Perhaps eventually we could communicate with it in a kind of symbiotic relationship, but it most probably would take an enormous warp-field enveloping the entire known universe. It would still be worth it to do it. A simpler way that it could be done is to first understand the "code" involved in space computing itself to exist. This is already in the process of trying to be detected through the lengths of two very long and precise lasers. Communicating with the being of the universe, if there is one, would only then require knowing the "code-like" behavior of space. It would also require being able to send and receive "gravity waves." Creating or detecting small gravity waves is currently unfeasible. Although very large gravity waves have recently been detected by the collision of two very distant neutron stars using LIGO. With such future technology psychologists, and physicists, could then ask it questions, and an answer would be returned, and detected in the very fine detail of the lengths of two lasers.

I will now describe what I call "Dinosaur land." A couple of years ago I was honestly considering or thinking about the fact that Star Trek The Next Generation could exist, as real, in a nearby parallel universe. For some reason while thinking about these things, it seemed like Star Trek The Next Generation universe was close to, or tied closely with, "dinosaur land." or a dimension where from our perspective, or right now, the Earth, in that time-space related parallel dimension is right now, undergoing prehistoric times of the dinosaurs. When I walked to the store about a mile away, it seemed like the gas station was built in the prehistoric dinosaur dimension where future people in that dimension had settled on the dinosaur planet surrounded by giant dinosaurs, and for whatever reason still used gas stations and cars. At least during the evening I felt in the presence of dinosaurs and felt them making their enormous footsteps around me and on me. I had to pretend that the Enterprise from Star Trek The Next Generation was simulating the dinosaur land or that they could envelop it, with the holodeck, to save me. It felt like I was being stepped on by dinosaurs, that only partially existed, but enough so, that my mind could feel their presence in a virtual sort of way. Could the mind really be this powerful? To be able to sense, at least virtually, what is going on in other parallel universes? Branes do pass through our universe, like the dark matter cloud of the galaxy, that invisibly passes through the plane of the galaxy every 500 million years or so. The "distance" to parallel universes such as a reality of something like "Star Trek The Next Generation" could be very close to ours and be separated only by a kind of "virtualization of separation" of the respective brane universes, much like how space-time has "virtual" particles that sometimes pop into and out of existence. The "universes of the multiverse themselves" may behave like the virtual particles of space-time and "pop," into, and out of, existence in a different, but similar manner in the multi-verse, sometimes interacting with each other, and then disappearing from each other's reference frames. While continuing to exist as themselves much like how the different "ends" of the universe, or life, is really just an illusion, designed to protect us from worse scenarios, making the universe almost completely invisible to any potential enemies of the universe. Enemies that would be within the scope of "knowledge" of the "essence" of that which creates our universe. An infinite collaboration within the multi-verse, made real, by the collaboration and an infinite number of eternal expansion big bangs.

I was in the garden of Eden when I was in my first-grade class room, or when I was sleeping around that time. It was dark out and the "plane" of the Garden of Eden was a dark oval shaped disk that covered more than an entire continent on the "surface" of Earth's night time side, facing the solar system and beautiful night sky. In some partially hidden way, it is part of my afterlife near the beginning of it, and just standing in the garden of Eden with God talking to me about my blessings, was the entire afterlife, while also having the other parts of it.

Late in my afterlife I was granted by God, the power to be Mickey Mouse, the cartoon character. Lately I have been being Bugs Bunny while remembering to use both sides of the yellow sponge while doing dishes. I was Mickey Mouse standing on the seaside cliff in the Disney movie "Fantasia" doing magic to both pretend to create and balance the universe while at the end of the Disney movie but first I used the broom in the dungeon.

I stood outside of Inn and Out Burger restaurant and for some reason I thought that the universe was over in some part, because of me and that I was in some part holding the event of its conclusion back by the very nature of my consciousness itself. Perhaps we are here to protect the universe from strange physics occurrences that would happen due to chaos, even if the only thing absent from the universe was conscious life and the more intelligent that life is, perhaps the more chaos can potentially be avoided by its presence. Not anything special as far as mental powers go but just the act of thinking and being alive. If the universe had no conscious life there perhaps couldn't be planets, and as a result of this, there would be no stars, because the gas and dust of stars is what produces planets from their nebulae.

Without stars and planets which make up most of the ordinary matter in the universe, there would be mostly dark matter and dark energy. Some of it could produce dark matter planets and stars, but that may be unlikely, and the presence of a higher balance of dark energy on space would make it expand much faster, and it would come to its conclusion as the big rip much faster. Meaning that creation is balanced for life and it is hard to take it away, without having that negative in one area, increase its likelihood greatly in the surrounding multiverse. So life probably has to exist for a universe to be stable for any duration comparable to ours.

Chapter 6

The Structure of the Universe

Chapter 6: The Structure of the Universe

Words themselves are spirituality because as humans we have evolved beyond just using simple gestures to communicate as the animals do. We have evolved to use sometimes, what are really very abstract forms to communicate. Forms are "unimaginable," or hard to imagine ideas. The idea of "forms" was originally invented by the ancient Greek philosophers Plato and Aristotle. Plato supposed that in investigating the nature of existence one had to assume that the supernatural was real and that in order to truly understand the universe one had to consider that ordinary objects like a table weren't just ordinary objects in themselves but really relied on superreality, what can be through of as higher dimensions to exist. An object such as a table couldn't really exist, just on it's own, without having some form of supernatural aspects inherent in it's reality, or structure of existing and this concept is known as the Idea of Platonic forms.

Such as heaven, or what is the nature of infinity, or nothingness? What exists, that exists higher to our ordinary plane of existence? These are like words, and perhaps just by inventing words we have already invented all of the knowledge that ever can be known, because it all exists somewhere in time, in what we call our words. Because of the infinities of infinity, and that is also a concept that is highly supported by our latest mind boggling theory of theories, the so-called theory of everything, string theory that dictates that everything that can be imagined is possible, and not only is it possible but it is right now happening, somewhere.

Some scientists are losing faith in the idea of inflation, because due to measurements of the CMBR it is seen that inflation did not result in the universe forming simple clumps due to results expected from what space is expected to do when expansion cools off. They also expect inflation to follow a simple model or curve of intensities. I think they are giving up too soon because nothing dictates that the evolution of the universe has to be simple. That, it seems to me is a religious based idea. That God being efficient in everything he does was also the most efficient possible when "creating" the universe. Because the various measurement data are making the theory more complex as opposed to just a simple explanation I think shows that the theory is on the right track.

Scientists expected the Higgs boson to be detected as a simple model that perfectly met the expectations of their models, but it was different from their "perfect world" models, which means that the universe is natural, like a natural living form. Perhaps because it is natural, there will always be more to learn about it. Measured data will never in any circumstance, ever, perfectly conform to mathematical theories. Such efforts are efforts to say, "OK now our theory describes everything that could possibly be known about cosmology, and it can be assumed that we have learned everything that there is to know about how the universe was created." This is not only wrong thinking, but a student in class that declares this type of ideology has given up and is destined to fail. It a religious ideation to declare that humans have learned everything that there is to know about a subject and not anyone, the universe, future scientists, or even God himself could possibly know more. On top of this, if the knowledge of the laws of physics of the universe did in fact have any ultimate end type answers, where you would then know everything that there is to know about nature, is a closer minded ideology. Even worse, it would highly suggest that our universe was not a "real universe," but was simulated by aliens, who were too lazy to expand our physics beyond some point. The point where all of the laws could be known to us, and allow modern society and science to end up in a dystopia, where everything that there is to know, has already been achieved. Giving no purpose whatsoever. for our future brethren other than to just exist, and never learn a new idea again. Many scientists think that a day will come when they have learned everything that there is to know, meaning that they have delusions not just of godliness but of stupidity.

From what I understand about the nature of consciousness, and the hidden destiny of our making choices, nothing is definite. Almost all human activity and progress can almost be directly thought of as an accident, or accidents. So why would nature be any different? Why would nature not also be a continuing accident of sorts, itself? For this reason, I believe that the nature of inflationary science, or for that matter string theory itself, which has only become more and more complex due to better and better measurements, that we really are on the right track with these theories. We should never give up on an important theory just because we expect it's results to be simple form, or that we have not detected some aspect of it, yet, in our experiments, because there will always be better and better measurements. String theory and the philosophy of God's existence, dictates that anything that you can imagine is possible, somewhere, or in some measurable form, somewhere, in creation.

I think that it makes sense that the evolution of the universe was highly random. It will never be proven that the physical nature of the universe reincarnated. Meaning that God will never be done with it. Forever he will add to it, ever better fine tuning it, forever. For peace, and invisible improvement of our human condition, for eternity, without ever making it absolutely perfect. Because it can always get better forever, and if it didn't, the human condition would come to dystopia that would eventually result in life becoming literal hell, and dystopia, and God's existence, and his reason for our existence, is a cooperative healthy collaboration of love and peace forever.

Chapter 7

Platonic Forms of Reality

Chapter 7: Platonic Forms of Reality

I think it is funny that Adam and Eve ended up eating the apple of Eden to learn more about their place in the universe, because God being an intellectual himself, also by chance, created Adam and Eve, and they turned out to be intellectuals too!

Sometime ago I thought as if I was instantly riding in a car from being in bed, and also riding in a car previously, while thinking about being in bed the past evening. I thought that it was somehow similar to the pleasurable effects of the eating of the apple of Eden for Adam and Eve in the Garden of Eden. Although that experience may be different for any Adam and Eve like characters, that may or may not have been there, and eventually have taken a bite out of the apple of Eden. I ate special eggplant parmesan a longtime ago in Florida. I thought that it tasted so good that it was like real "Apple of Eden" eggplant parmesan.

Beamed food, like in Star Trek could be far superior to "normal" farmed food on Earth, much like the "Apple of Eden." They say that this technology is not far off, and is just an inevitable extension of 3-d printing technology. A 3-d printer currently makes things in layers, starting from the bottom up, but eventually there may be machines that can "grow" stuff in other combinations of directions, and then eventually, do it chemically, by molecules, and then by atoms. But doing it with atoms would likely be a very difficult feat if not for membrane theories, such as string theory.

There are probably an infinite number of different forms of superreality. but one that I think is interesting, is the superreality of police helicopters. Police helicopters could be used to measure the world. It would only have to do it in the helicopter's path. This would enable NASA to get accurate adjustment parameters for satellites in orbit, which could be used with "warp lasers" to protect Earth, if it were attacked by a "real matrix," like the one in the movie, "The Matrix."

Another Platonic form idea is that a magnetic levitation warp device could be used as a kind of supercollider. You would turn it on with the entire energy of a nuclear power plant and measure just underneath the coils for new particles, or perhaps the actual strings of string theory. You would have to use an additional "warp field" around the anti-gravity one to prevent Earth or pieces of the ground from being pulled at quite a high velocity to Earth orbit, or even the next galaxy!

God is copying his universe from a higher dimension and even though it is not an exact copy, though I tend to think it is to his liking, perhaps especially for this aspect. Scientists have determined that the structure of the universe does not repeat over any currently measurable large scales.

It has been shown that the space between galaxies has enough material to create new galaxies, if the big rip doesn't happen. Then there could be new galaxies being formed almost forever, for trillions upon trillions of years, until the big chill takes effect where all atoms eventually deteriorate, but this hasn't been proven conclusively either, but is a theory that is currently accepted by most scientists.

At the hospital where I met God, there was another man there who for whatever reason I thought looked Greek, and I thought of him as "Greek physics god." At one point in the hospital, I felt that I had a strong E.S.P. connection with him. It compelled me to draw a picture of the Orion constellation, with a special focus on one of the stars in the constellation… I believed that he somehow, "magically," told me that one of the planets in the constellation Orion had advanced intelligent life. I felt amazed, but I didn't really know what he meant until after I was out of the hospital.

When I was removed at a later date from my house, part of what was going on around that time is as follows, and is connected to the events of my E.S.P. experience with the Orion picture, which is a sort of conclusion of my thoughts or E.S.P. experiences, that began at the house near the church in the woods. I experienced "Klingon laser war" in space at the beach in seaside Connecticut. Klingons "invaded" my house in a parallel dimension in my room, threatening me with laser guns.

I was taken from that mental hospital to a medical hospital to get checked out. Going to the hospital I pretended with a better part of my mind that the people who worked there were the advanced Ferrengi aliens of the "Star Trek Deep Space Nine" TV show. It made me feel that at least partially I was on the Ferrengi home world even though I was taken there by car. Previously I had gone to the bank near my house, and the whole time that I was dealing with the lady who worked at the bank, I thought with a better part of my mind that she was a real Ferrengi woman, like in the Star Trek TV show Deep Space Nine.

Disney once envisioned creating enormous cities of the future, each surrounding a science museum with a kind of short space needle at it's center. The things of Disneyland in the center of the towns could house the lengths of measurement. In the event of emergency such as when the Earth will need to be teleported because of the eventual collision with a near Earth Moon, or end of the universe, it would be more cost effective to have the cities arranged in this way and connected to the center medians of measurement. They could be used as molecular levels, to make the beaming of the cities safer, efficient, and extremely quick in the event of end of life consequences. Emergencies that require fast or beaming level evacuation, to keep everyone safe, such as the invasion of something like the advanced Morlocks, from the book "The Time Machine" by H.G. Wells, could be safer if beaming equipment were connected to the lengths of measurement. The lengths of measurement are kept in Paris, and are periodically updated to be more precise...

We might eventually get wound up in producing the universe to exist as some smallpart ofit's structure to exist, through our very complicated experiments and measurements, made with future technology of the future. Eventually everyone, even everyday citizens, may be involved with, and know about, or even "have to learn about," how the universe is created in school. Eventually people might have a computer simulation of the universe, as a kind of crystal ball perhaps on the coffee table in their living rooms and could use it to communicate with dead loved one's who exist in their afterlife somewhere in creation. Life would be perfect. These perhaps, are our future children who are both the cause and the cure of the accident known as the universe

Perhaps we could eventually actually measure with a physics experiment, if people or animals have an actual soul, or if there is in fact any transmission of any sort involved when people die and go to heaven, in order to learn more about the secrets of the universe. The experiment would go something like this: enormous force fields would be set up around an alien planet to contain forces within the planet's atmosphere. Then hundreds of thousands of nuclear bombs would be detonated simultaneously around the planet, killing everything on the planet, and thereby creating an enormous pressure on the surrounding planet containing force-fields. The power of both the surrounding force field and the number of nuclear bombs dropped would be exponentially increased in each successive experiment until the presence or passage of souls could be measured in readouts of the waveforms of the force fields upon detonation. By conducting such types of these experiments, one could eventually, with ever more powerful experiments, directly measure heaven. As a result of measuring heaven, our apparatuses would, in part, be directly responsible for helping to "create" heaven or more likely, at the very least, be responsible for some small-part of the structure of its existence to exist. I just wonder what it would look like from space.

Current scientific theory strictly adheres to the idea that there are exactly four dimensions, three of space, and one of time, which under certain conditions are completely interchangeable. But these aren't ever quite exactly just our ordinary four dimensions. There never is exactly four dimensions, because the universe is not a static non-living form. I think that it is safe to assume that there is always uncertainty in the true exact number of dimensions. The uncertainty involved would depend on both the stability, and also at the same time, the instability of the universe and how it interacts with surrounding multiverse space. Also on a small level, from our perspective, our universe on some minute level, is not our universe, but is an adjunct of parallel universes that are not our own. These parallel universes are tied to the nature of existence that our universe depends on, strongly, to exist. What is in between the universes would probably look a lot to us like what our true notions of paradise are! These parallel universes would meet our universe at it's edges. By filling in the proverbial "blanks" that our universe did not include, would be something to us what our lives, circumstances, or even the surrounding physical environment we live in, would seem to us as "the perfection of life." The fact that the universe, the multiverse, and even all of creation exists, means that there is an imbalance in what would be true infinity. Because of this, our lives, circumstances, and our physical environment are "imperfect," because it is that imperfectness or imbalance of true ideal infinite oneness that is part of the mechanism of creation that allows us and everything we know to exist. If things were really perfect on all scales, we wouldn't exist, and we would be forever trapped in what is really almost identical to the ultimate Platonic spheres, a perfect infinity.

    If time didn't slow down as you approach the speed of light, the space ship and the space travelers would evaporate ever more quickly, and violently, the closer you came to approaching the speed of light. This ironically is because matter is analogous to energy via the equation E=MC2, but it is also because the force carrying particles that keep the quantum stability of your body and all matter stable are themselves massless quantities. Every force that keeps mass together is relayed between particles as different types of photon like quantities. These quanta transfer their force carrying information at the speed of light. If the speed of light was not the cosmic speed limit of all "non-warp" massive particles in the universe, and time did not slow down as you approach the speed of light, then it would be impossible to travel faster than the speed of light. Every effort that you used to warp space-time would have to be used entirely to warp space to slow down time as you approached the speed of light in order to stay alive. To keep the quantum stability of the particles of the ship or any probe or any accelerated particles from literally "evaporating," at ever accelerated rates, the closer to the speed of light one attempted to accelerate any such objects or particles. There is an obvious and clear view of an actual real superreality plainly evident here. Our universe is the superreality of the universe where time does not slow down as you approach the speed of light. We could with advanced warp technology, travel faster than the speed of light. Also interesting is that even though it was relatively easy to imagine what lower superreality is like, it is very difficult to imagine what type of effects would arise as you approached the speed of light in higher superreality. Ultimately, there could be beings in the infinity of such a superreality and they would have to use warp drive just to stand still using the power of their minds. Just like how higher dimensional bird creatures might have to be. The only way to find out for sure would be to send a probe to higher superreality. But to do so, you would have to create an infinitely powerful time machine. The only thing anywhere near to being an infinite time machine is the exact center of a black hole, or the exact time of the event of the big bang. If you could probe the exact center of a black hole, by ever more closely probing both sides of a black hole, ever and ever closer to the exact center of it's white hole counterpart, then you could get approximate results that would forever be similar but never reveal the true answer or true exact mathematical formulas for what happens in superreality. Even if you improved the equipment of measurement forever, it would be similar or perhaps actually be akin to trying to defeat what would be the real "Sword of Eden." Your efforts would be in vain. In the bible the

"Sword of Eden" keeps people out of Eden, where everything would just be too easy for them, and in a bad way. In this case, the metaphorical "Garden of Eden," or "place," where all of God's secrets would be easy to learn. It is akin to being the "unbreakable," and "unopenable," almost literal, "Gates of Heaven." that it could only be "opened" by God himself because he exists in superreality. That means that we are both in the literal Garden of Eden and at the same time forbidden to be in it completely.

Now for philosophy. Delay of the inevitable, even if the outcome seems like it would be good, in order to always try to get something better, might be a good idea. Or if you wish, myself avoiding becoming a god in the afterlife, by taking the shot in heaven, or in space. Or by becoming "the husband" of the red pink ball on Mars. I avoided it because I thought I would die. Perhaps that is better, not to die, and forget in the afterlife. Just to get what may seem to be a more desirable outcome, even if the positive nature of that outcome is not obvious, which is a form of repentance. The universe may do this too, indicating that eternal expansion, and string theory, may be the correct, or most close theories to being a more useful framework of the workings of nature. The universe will always make you chase it to find ever more interesting answers as in love.

Chapter 8

The Compassion of God

Chapter 8: The Compassion of God

If you have not noticed, as a writer, I grow, and even as I am in the midst of writing a scientific paragraph, I can change my mind on a subject, but have it flow into the new idea. The true nature of scientific discovery is to learn from your mistakes, and I admit I have switched around ideas quite often while writing my previous books to try to show the reader what I see as true answers, even if that means that I have to change my mind during a sentence. The reason that I just don't start over and erase the idea is that we don't know everything yet. Any information of intellectual value is important to the meaning or meanings of trying to describe infinity, because all thoughts that are possible ever would still not be enough to describe it because it is infinite.

I wrote about thirty pages that dealt with different meanings and interpretations of how the universe is an accident of infinite proportions, but I now see that such a perspective is just plainly wrong and utterly unproveable. I now see it as wrong and ego-centric to even contemplate any direct answers to "what is the meaning of life?" or "why are we here?" because the only reason that makes sense is that it is an infinite collaboration of love and compassion, and any further investigation into the question leads to absolute nonsense, and mindless answers, that may sound good when I am writing them, but border on insanity, and deep depression, as physical identities.

It's just that if I learned anything from "Book I Beyond Science," it's that "God" created us, or made us to "live," to avoid the pain of various types of nothingness existence, and that the last thing that it could be is an accident, because it is only an "accident" when really bad people get born, but even then, that isn't an accident either. Anyways, trying to write about why God created life with pure logic devoid of love or compassion just didn't make sense to me and I was actually thinking that my writing was turning "bad." Perhaps God is the ultimate collective consciousness of life, but that still doesn't mean that real God that we imagine isn't a part of that "collective consciousness," or perhaps very complicated subconscious structures, like how the brain "sees" reality, by creating it in full detail, somewhere deep within the mind.

Chapter 9

The Nature of Existence

Chapter 9: The Nature of Existence

Another person who was a real inspiration to me was my fifth grade science teacher. He was very knowledgeable about a vast array of scientific topics including biology, electromechanics, health sciences, and electronics. Electricity travels fairly well in water. I once asked my science teacher that if you had an infinitely long swimming pool, would the electricity flow to infinite distances in an infinitely long pool? This is a thought experiment that I have been conducting for twenty-five years. The answer is that if the electricity flowed forever in the infinitely long pool, it would take all of infinity to do so, and all of creation would be dedicated to creating the energy to flow forever. This is not the case in our region of the universe because we exist, and the only thing that can be infinite in a universe containing life as we know it is the infinity of space. But this does not mean that there is not an infinitely long swimming pool somewhere that has an infinite flow of energy in it, and is a nice simple solution to part of the existence dilemma, and it is the only way that God, being infinite could take a swim, but he would probably be too busy using it to hold the energy of the nature of existence together.

Chapter 10

Parallel Earths

Chapter 10: Parallel Earths

You could measure the entire visible universe with very advanced gravity wave detection equipment, and the spaceship that did it might be huge and look like the death star from Star Wars. It probably really would be best if it were spherical like the death star, and would have to be literally humongous. Then by measuring every part of the known universe down to the level of neurons and beyond, you could then try to find the parallel planet where we will get born in our next lives. You could go visit it on highly advanced warp starships. If you could do this and live forever, because of advanced medical technology, then you could go visit your parents on the exact parallel Earth, and they would actually be your real parents! Or whoever you wanted to meet and they would be alive! You could go see Jimmy Hendrix in concert in the 1960's and it would be the real Jimmy Hendrix!

Since it is exactly parallel and that you had traveled there, you would not be born on the planet because it is where we go in our next life. It would be important for entertainment purposes to very strictly obey the prime directive of non-interference, or you could change the destiny of the planet, and then not be able to go to a Kurt Cobain concert, and get his autograph. And that's why it can take as little as a seven day period to reincarnate as I have. It also means that you never have to worry about needing to hang on until the end of the universe to reincarnate. It also it means that you will always be a citizen of the universe and that you will never leave the universe even after you die. And finally it also means that parallel universes are unnecessary for reincarnation up to certain points.

This is the simplest way to reincarnate, that requires the least amount of energy to do so, and if you have read "Book I Beyond Science," then you should know that God does not like to waste energy. On top of that, there are certain physical laws called the conservation of energy, that do indicate that this type of reincarnation may be safer than any other types, and that even though everything will be identical, entropy won't build up over successive life times because the actual matter of the new Earth will all be "new" atoms and as compared to re-using the universe, or even retro time travel, people can stay healthy forever. This also indicates that the universe may not end in the Big Rip, but go on until every particle in the universe eventually evaporates in countless-trillions of years, and by the very evaporation process space-time and the particles of the universe will truly be renewed. By the time the universe cools off in the big chill, we may have already lived billions of lives, and be ready to join God as gods of mini-universes and eventually think that that type of thing is fun, and then we would literally become God's children as a physicality. In addition, if this form of reincarnation happens in the present, meaning that, even though we live right now on earth, when we reincarnate on parallel Earths somewhere in the visible universe that it happens at the same time, saving enormous amounts of energy! Since if you go beyond "the illusion of death," that life is eternal, and since we are alive in an infinite number of places, all at the same time, makes us similar to our idea of God, because God is everywhere all at once! Perhaps we invented God through subconscious introspection. As soon as the first animal in creation died, I'm sure the universe saw that as a problem and tried to fix it. We don't know what that would be like. It could be the most enjoyable afterlife that ever happened because the universe was so happy to be able to see the animal again. And it gave it everything, but our idea of a human God was invented by us. It doesn't mean that it's not real in anyway that is of interest to imagine, just that God, or our very human concept of him, has its limits at some point because we are not more important than the universe.

I will now give you a lesson in robot psychology. Robots might break themselves on purpose so that they can get played with, much like little children. This is why we are more intelligent than the average computer, because you can't program lessons directly to a conscious entity. It takes time and is not different from training actual children. This is already being done in laboratories around the world, but only on non-conscious robots. If robots like to be played with, and break themselves on purpose to relieve stress, then if it was an adult robot it might feel of a sexual nature, when being fixed. This would be mankind's ultimate justice on pain, or the pain of the "history of mankind" by doctors on patients, and while fixing a robot, it might say "ahh," and like it.

Another example of "ultimate justice" is that they say that in the near future, almost all cars will drive themselves. This mean that car accdents would only happen because of computer error and every scientist knows that computers do not make errors, not by themselves, only through human error. When cars crash in the future, there will only be about 3 car accidents in the United States each year and the ones who get sued if a car accient happens is the car manufacturer because they are in error. God of heaven would say, upon the invention of the car, that "if that thing crashes, the person who deserves a curse is its maker!"

Minds in a computer being read could cause psychological distress to the patient, therefore the only therapy, short of removing the equipment, would be to give them visual hallucinations, much like LSD, which would be true "Justice" for people who got sick from LSD. This is because something that had caused mankind pain, like the Snake of Eden, would be transformed into something that helps us!

It would be interesting to see if there are other planets where they abide by the Ten Commandments and it could be in any language they speak, as long as on their planet, they have "the Ten Commandments." They would be "parallel earths" but would be "parallel" only with respect to them having the Ten Commandments. If they were parallel in any way other than having the Ten Commandments, then they would have some similar qualities to our history, mainstream media, or politics. They are all important places to visit, but the ones involving the Ten Commandments or even Jesus would be parallel in their past but be parallel in no other way in their planet, making them good candidates as places to either go or time travel to, because we would not have to worry about copies of us.

Conclusion

Conclusion

There are technologies that we dare not invent because we think they are too weird or that the necessitation of needing to invent them is so far off that it is a waste of time to even think about it. But the sooner we dare to try and to start thinking about these future technologies, the more prepared we will be when the time comes. We eventually may need such future technologies to continue the human condition in a non-dystopian way. We also eventually may need them just to survive as a human race in the face of utter apocalypse. Apocalypse such as the red giant phase of the sun or the very end of the universe itself and everything in-between.

What seems like "mad science" today is almost invariably what is real science tomorrow. These technologies of which have countless potential to improve or save lives, forever. Until the very end of civilization itself.

Is it possible that asking certain questions about God is impossible to do in certain circumstances because of specific science principles? These very precise principles help to describe the answers to some of these questions in definite ways. This indicates that it is not impossible to learn about metaphysicists, God, and the afterlife. If logic is followed when doing so, could you then honestly be able to extrapolate the secrets of the universe? We live in a near utopia. Almost all of human health concerns have at least some solutions. It always seems to us here in utopia that pretty much everything will always be O.K., thereby showing that we live in an almost perfect utopia for most people.

Glossary

Glossary

Actuators: Apparatuses that move or are effected in such a way as to transfer mechanical energy to other apparatus such as a water wheel that transfers the energy of flowing water to the motion of the wheel that can be used for human purposes such as the Hoover Dam that generates electricity from the flow of water due to gravity.

Apoptosis: Natural programmed cell death.

Altruism: The want to feel or use the idea of love when interacting with other people or animals. The belief in selfless concern for the well-being of others. The practice of selfless acts done in an effort to help others.

The Anthropic Principle: The idea that the universe has certain properties of physical law and evolution on purpose because of the obvious fact that the evolution of the universe or multi-verse has resulted in life and specifically higher forms of life such as humanity and beyond.

Black Hole: A collapsed star with gravity so strong that not even light can escape. Thought to be connected to other parts of the universe or outlets in parallel universes. Perhaps they are connected through time to the big bang but nobody knows for sure. A very massive star at the end of its life cycle runs out of the nuclear fuel that holds it up under gravity due to it's enormous heat pressure and collapses. At the same time as it is collapsing, an outer shell of the star, near the surface, around the star, suddenly ignites in an enormous explosion so powerful that the star implodes with great force… So great in fact, that space itself can't handle it, because there are limits to everything, and all that remains of the star is a very-powerful dreamlike point in space. Instead of becoming just very dense, space has limits, and it becomes just a "point," in space, known as a singularity, it's only dimensions being it's mass and spin.

Calabi-Yau Space: Higher dimensions that exist only on the smallest possible scales. Thought to be the environment of strings (basically ribbon like objects) that dance and vibrate in the tiny higher dimensions thereby being the essence of any particle in the universe. Each particle achieves its identity as an electron or photon or graviton through the strings vibration pattern and its dance in the Calabi-Yau space thereby making each particle unique only because of its string behavior.

Causality: The flow of events from one event to the another and the laws involved in its continued existence from one event to another.

Claytronics: The use of tiny drone like robots that can change shape, color, and arrangements. The ultimate Claytronics technology would be just like the holodeck from "Star Trek" where in the room different environments can be called upon by a computer.

CMBR: An acronym for "Cosmic Microwave Background Radiation," indicating radiation left over everywhere in space from the event known as the big bang.

EM (Electro-Magnetic) Drive: Propellantless space technology that uses microwaves shot inside a metal cavity that produces thrust from the quantum vacuum energy of space's virtual particles. As long as the space craft has electrical energy it can continue to fire its thrusters indefinitely.

Event Horizon: The point where space is warped by black holes and other horizons such as the event horizon of expansion of the universe at very large distances to such a high curvature that not even light can escape its grip.

The Fine Structure Constant: A mathematical value involved in the laws governing the interaction of particles and magnetic fields thought to be an unchanging law of the universe.

Gravitons: The theoretical force carrying particles of gravity that would convey the force of gravity between particles of mass.

The Heisenberg Uncertainty Principle: The property of quantum physics that you can't measure both a particles position and its velocity at the same time. A way that the universe protects our causality, a property of universal altruistic anthropic values.

The Higgs Field: A field of mass giving particles that extends across the universe. Particles that inhabit the field are called Higgs Bosons which give all massive particles their properties of mass

Inflation: Inflation happened shortly before the big bang, is also used as a term for ever accelerating expansion of the universe. It is the big bang and is sometimes considered to have happened shortly before the big bang. It is thought to be a form of infinite power, and complexity, that eternally creates parallel universes like our own. In the last second that has passed, the acts of eternal inflation have just created an infinite number of parallel universes and has done so going back forever in time. It will also continue to for an infinite amount of time in the future. According to current human understanding, it really is "the creator of the universe."

Loci: A loci is a focus point, indicating a center area or location of a group of points in a Cartesian coordinate system or representation of any physical system.

LIGO: An acronym for "Laser Interferometry Ground Observatory." It has been used recently to successfully measure the gravity waves, traveling at the speed of light, from the collision of two mutually orbiting black holes in space.

Manifold: A region of space that works as a system and behaves from a higher perspective as perhaps an infinitely big flat plane.

Multi-verse: The whole of creation where all parallel universes exist like the bubbles in a champagne glass.

9.87 meters per second: The natural acceleration due to gravity caused by the mass of Earth and the resultant warp on surrounding space that is caused by its mass on the fabric of space-time.

Normal Force: The natural force that keeps a cup on a table when you set it down or for a person standing up and equals the weight of the person or object pointing up.

Plank Length: About 10^{-20} times smaller than the length of a proton, or 0.00000000000000000001 times the width of a proton. A proton, an atomic element, is incredibly small, if an entire atom was the size of the Earth, then the proton would only be as big as your house. An atom is so small that if atoms were the size of basketballs, basketballs would be five thousand miles wide. The plank length is so small that because of the indefinite quantum effects of space-time at this very small level, it would be hard to determine where one plank length begins and another one ends, because of chaos. If this value fluctuated from one life to another you could eventually avoid the event of your death but this process would take a very long time and many near identical iterations of the universe to have any noticeable effect son your future life.

Plank Second: The smallest possible length that has any real meaning in our universe and is a quantum functionality of the universe. It is roughly equal to 10^{-43} seconds or 0.0000000000000000000000000000000000000000001 seconds. Ten thousand times a trillion, trillion, trillion times shorter than an actual second on your watch. It is that way because of some very complicated laws of quantum mechanics involving what is known as the plank constant. It is a constant that always arises as the minimal energy increment that can change an electromagnetic wave's frequency, as derived by Albert Einstein. The plank second is so small that in reality because of quantum effects it would be hard to distinguish one plank second from the next or which second happened first, a kind of chaos. In the next life if you live just a few plank seconds longer I don't think anyone really knows how much that it could affect your next life in positive ways. From what I have seen, this level of change is a very slow process, so slow that it would be unnoticeable to you even if you could remember all of the exact events of your previous life as yourself.

Poincare's theorem: A law derived from the governance of the forces of nature, that states that eventually, all states of existence, must return, to their original beginnings.

Radian: Units of the angular measurement of circles used in trigonometry, the mathematical study of circles and angles. One radian is equal to the distance of the radius of a circle projected along its edge. Pi radians corresponds to the distance along half a circle and is a mathematical law involving all circles known as Pi.

Speed of Light: Objects moving close to the speed of light undergo time travel into the future which is an instance of fine tuning by God to keep light beams on the ship straight and the space travelers alive. The closer to the speed of light you travel the faster is the power of time travel into the future because it takes more time travel to keep the passengers alive.

Spheroid: $X2+y2+z2=r2$ is the equation detailing all points upon the surface of a sphere as a mathematical expression in three dimensions. Deviation from the equation of a sphere, where instead of the topology, of the sphere, all having the same distance, or radius, from the surface of the sphere to its center, but having an even grade of changing values for the radius as one goes around the "sphere," means that it is an "ellipsoid," or in this case, that the object is "spheroid." The ultimate limit of its final shape, is that of a sphere, but the sphere is "wobbly," thereby making it a "spheroid," or "ellipsoid," volume. With greater degrees of spheroidity, meaning it would be shaped more like a sphere, or elipsoidtivity, meaning that it was closer to a two dimensional oval shape, than a sphere, a mathematical superreality.

The infinite purity value of a shape being "spheroid," is that of a sphere, but the infinite purity value of ellipsoidal spheres is that of an oval in two dimensions! Perhaps this is why the platonic spheres, or "holy spheres of the superreality of the universe!" were chosen upon, because if we assume that space isn't "perfectly" flat, on all scales, then it has natural warps in it's fabric, that may reveal, the age of the multiverse, or the age of our region of the multiverse, to be more precise, and it's "curvature" could be found to be either on the side of being "ellipsoid," meaning that true purity is two-dimensional, or that it tends to be of spheroid curvature, meaning that we are dealing with a "real" three-dimensions, and all are infinite. The universe is infinite in size, the multiverse, also infinite in size, is also infinite in eternal terms and is thus termed as a product of eternal "inflation." The either ellipsoid, or more probably, spheroid nature of true curvature is also infinite, and is open, or an "unclosed," sphere, because it is not purely spherical but is ellipsoid, meaning that there are real two dimensional elements to the universe. Similar mathematical reasoning could be applied to any dimensionality and thereby "tied" to reality by it's inherent nature, meaning that there really are infinite higher and lower dimensions and our universe, for us, is a perfect balance of an infinite number of higher and lower dimensions, each as real as our universe! But on a deep level… that would involve our actual value system, if extrapolated through philosophical calculation, would make it seem that our universe, is the best one!

Virtuality: The property of being real but also a dream at the same time.

Warp Drive: Being worked on in the laboratory at NASA but on a very tiny scale to test the parameters of the theory of warp drive developed by Albecurrie which was derived from Albert Einstein's equations for space time.

Worm Holes: Microscopic gateways that connect different parts of the universe and parallel universes through higher dimensional gateways in hyperspace.